CANYON AND COSMOS

Canyon
&
Cosmos

Searching for Human Identity
in the Grand Canyon

DON LAGO

UNIVERSITY OF NEVADA PRESS | *Reno & Las Vegas*

University of Nevada Press | Reno, Nevada 89557 USA
www.unpress.nevada.edu
Copyright © 2025 by University of Nevada Press
All rights reserved
Manufactured in the United States of America

First Printing
Cover design by Elke Barter
Cover art by Serena Supplee

Library of Congress Cataloging-in-Publication Data available upon request.

ISBN 978-1-64779-195-7 (paper)
ISBN 978-1-64779-196-4 (ebook)
LCCN: 2024038705

The paper used in this book meets the requirements of American National Standard for Information Sciences—Permanence of Paper for Printed Library Materials, ANSI/NISO Z39.48-1992 (R2002).

Contents

Introduction 1

Prologue: The Sipapu 5

The River 17

The Abyss 133

The Raven 247

Acknowledgments 351

About the Author 353

CANYON AND COSMOS

Introduction

When Henry David Thoreau moved from town to Walden Pond in 1845 to test out his belief that nature is more important and meaningful than human society, nature was simpler than it would soon become.

Only weeks before Thoreau moved to Walden Pond, Lord Rosse in Ireland inaugurated the world's largest telescope, and he soon discovered "spiral nebulae." This was forty-four years before the birth of Edwin Hubble, who would prove that spiral nebulae were other galaxies and that all the galaxies were flying outward from some strange beginning.

The universe of atoms, too, remained unknown. 1845 saw the birth of Wilhelm Röntgen, the discoverer of X-rays, which hinted that matter contained strange energies. In 1845 Dmitri Mendeleev's periodic table of the elements was a quarter of a century in the future, and it would be a further generation before physicists mapped out atoms.

Thoreau had read Charles Lyell's *Principles of Geology* and was ready to see signs of an Earth far older and more dynamic than humans had imagined, but it would be another fifty-seven years before Alfred Wegener launched the idea of plate tectonics.

The living world, too, would change dramatically. In 1845 the concept that cells are the foundation of all living bodies was only a few years old. Charles Darwin's *Origin of Species* was fourteen years in the future.

At Walden Pond, Thoreau was not seeing an expanding universe, an intricate universe of atoms and cells, or an Earth with moving continents and evolutionary life, but his primary message would become ever truer and more important as science vastly expanded our concepts of nature. Our lives are embedded in a reality vastly larger than ourselves, a reality that gives humans deeper connections and meanings. Yet humans are intensely social animals and we derive our identities primarily from the masks we wear in human society. We devote enormous energies to trying to impress other people with our value, defining value by social roles, wealth, possessions, achievements, or appearances. Nature remains a mere scenic backdrop or economic resource, and science realities remain intellectual abstractions, barely connected with personal realities.

Along with his call to live as if nature is real and really ourselves, Thoreau gave us a powerful tool for exploring our connections with nature. *Walden* is not a scholarly presentation of facts but a personal exploration of what those facts mean for human life, how they offer us larger identities. Thoreau tested out ideas to discover how to live them, how they felt, how they changed our perceptions of the world. He was pleased to find nature's aesthetic beauty along the way, but he was searching for deeper beauties. Useful for his purposes was lyrical prose, which brings the world to fuller life—poetic symbols can carry a lot of power. *Walden* bequeathed to us a form of writing, literary nature writing, that has flourished ever since.

Yet literary nature writing has been slow to expand its horizons into the larger dimensions of nature revealed by two centuries of science. More recently, literary nature writing has been heavily preoccupied with environmental crises. Much of literary nature writing remains earthbound, ignoring the cosmos.

Humans have always needed to understand where we came from and how we fit into the cosmos. Yet the scientific cosmos remains more of an intellectual abstraction than a lived reality. We can give numbers to the Big Bang but we don't perceive its particles and energies pulsing within us. Literary nature writing can be a good tool for making science ideas more personal, letting us experience them more deeply. And speaking of deeply, there is no better terrain for this pursuit than the Grand Canyon, which embodies deep time, geological forces, and biological evolution, allowing you to touch

INTRODUCTION

them with your hands and be touched by them, to feel them within yourself. If you are paying attention, you might even experience cosmological depths here.

Paying attention requires you to go quietly, listening to nature and not your own ego, letting the rocks and the river and the stars tell their stories, tell you that their stories have always been your own story, that all along you have been primordial elements.

Another development Thoreau missed was the philosophical fallout from the discovery of the expanding universe. As science removed the traditional locations and roles of the gods, many people felt empty, homesick for a more spiritual universe. When Thoreau went to Walden Pond, Friedrich Nietzsche was a seven-month-old baby, and Albert Camus's *The Myth of Sisyphus* was a century away. Thoreau lived in the universe of Romanticism, which was the first major cultural attempt to answer the advance of science. If it was becoming harder to believe the book of Genesis, the Romantics would find God in the Book of Nature, which was full of divinity, a benevolent master plan, messages in rainbows, and moral lessons in the lives of sparrows. Romanticism remains a powerful cultural force today, still instructing our perceptions of nature. Even people who have stopped looking for divinity in nature continue repeating the postures of Romanticism, expecting pure harmony.

This book will also take a deep journey into what philosophers and poets called "the abyss." Once again, the Grand Canyon offers abundant raw materials. One canyon overlook is called "The Abyss," and it is one of the most popular places for people coming to the canyon to commit suicide.

Yet perhaps we can go even deeper than the abyss. If we are paying attention, we may be able to move beyond mere social identities, beyond the habits of Romanticism, beyond existential despair, beyond the science-given identities with which this book begins. We may find that the Grand Canyon, as a portal to primordial realities, is also a portal to the deepest primordial reality, the ultimate mystery of existence, and that this may be the deepest, most nourishing source of life.

3

Prologue
The Sipapu

THE SIPAPU

From here, you can see forever.

You can see deep into the earth and time and evolution. From here, rocks are not inert but full of power, and life is ancient and ever new, full of creativity. From the rim of the Grand Canyon, the earth is more real and the cosmos is closer. You are seeing your own depths, your own origins.

For the Hopis, whose ancestors lived in the canyon a thousand years ago, the canyon remains spiritually powerful, full of creation, the place where humans first emerged into this world from a series of underworlds. From the canyon rim at Desert View, I am gazing a dozen miles toward a side canyon that holds a mineral dome with surrealistic shapes and colors, from which golden water upwells and trickles into the oddly turquoise water of the Little Colorado River, a major tributary of the Colorado River. This dome is the Sipapu, the passageway from which humans emerged into this world and into which the souls of the dead journey to reach the afterlife. Hopi kivas hold a symbolic sipapu, a small hole in the ground, to remind them of the Sipapu in the Grand Canyon.

Two dozen miles down the Colorado River is another surrealistic mineral dome, Ribbon Falls, with a waterfall, sparkling with rainbows, splashing onto it. For the Zunis, whose ancestors lived here a thousand years ago, Ribbon Falls is their Sipapu. Both the Hopis and

Zunis make pilgrimages to their places of emergence and perform ceremonies to honor the forces and mysteries of creation.

For the Native Americans who lived in the canyon, it was not a brief tourist curiosity, and not just their home but their womb. They were born from its soil, they immersed their hands in its soil to shape it into corn and into their flesh, and they knew they would one day return their bodies to its soil. Their motions were powered by the motions of the Colorado River and the rain and the wind and the running deer. The red cliffs became their faces that at sunset glowed more truly red than sunsets ever would on white faces. For thousands of nights they watched the stars turning above the cliffs, and they turned them further, into stories and meanings. The canyon reached deeply into their mythological hunger and imagination. Agricultural peoples all over the world have turned their experience of how the world works, how life emerges from the nurturing earth, into creation stories with humans emerging from the earth. When the Hopis and Zunis attached this idea to the Sipapu and Ribbon Falls, as strange as creation, as deep inside the earth as humans could ever reach, they created the world's most dramatic symbols of humans emerging from the earth. No clay figurine of Mother Earth with a bulging womb could compare with this mile-deep, blood-red, river-fertilized womb. As symbols, the Sipapu and Ribbon Falls are endorsed by geology, for both are made of the dissolved bodies of ancient life, the eons and the life from which humans did indeed emerge.

One of the first geologists to contemplate the Grand Canyon, Clarence Dutton in the 1870s, also felt its mythic power. Dutton had studied for the ministry at the Yale School of Divinity but became a surveyor of the American West. When he saw the Grand Canyon, his religious impulses emerged and blended with his geological knowledge: he saw the forces of creation, a gash through the ordinary world and into cosmic mystery. He felt that the canyon would be insulted by the sort of names Americans were giving to frontier places, the names of pioneers and national heroes. Only the names of gods would do justice to it. To three prominent buttes he gave the names Brahma Temple, Vishnu Temple, and Shiva Temple, the names of the three aspects of Hindu divinity: the creator, the preserver, and the destroyer. Dutton named another butte for the Egyptian god of the underworld, Osiris. Dutton correctly recognized that the Grand Canyon would rouse people's deepest sense of wonder. Subsequent geologists, geographers, and writers continued Dutton's concept, giving canyon landmarks the names of dozens of gods until there were whole sections of landmarks devoted to Egyptian gods, Greek and Roman gods, Germanic and Nordic gods, Hindu gods, Chinese gods, and the Old Testament. Many of these names were given by scientists who felt little loyalty to religion yet who did feel the canyon's mythic power, for they were handling rocks in which they felt the magnitude of time, the power of geological forces, and the abundance and creativity of life. There isn't any section of Christian names, perhaps because many Christians were attacking geology as false. Yet John Wesley Powell, the Grand Canyon's first river explorer, who had rebelled against his father's desire for him to be a Methodist minister and who fully embraced the geological Earth and evolutionary life, named a sparkling creek "Bright Angel" from a new hymn that started "Shall we gather at the river." The same geologist, Francois Mathes, who named Apollo Temple, Solomon Temple, Walhalla Plateau, and the Inferno (from Dante) was feeling the same grandeur and mystery when he named Evolution Amphitheater, which is surrounded by evolution's discoverers: Darwin Plateau, Wallace Butte, and Huxley Terrace. Dante and Darwin were legitimate neighbors because both had looked deeply into the well of creation and the riddle of life and suffering and death.

Today many people come to the Grand Canyon expecting a visual spectacle or an extreme natural wonder, but some, even without being prompted by canyon landmarks named for gods, feel something deeper, something primordial. They won't see Ra or Jupiter at work there, for the human religious imagination has moved on and left many gods behind, but many people see the work of the monotheistic god, see the mind and hands described in the book of Genesis.

Science-minded people, too, see the powers of creation, for the Grand Canyon, better than any other natural wonder, makes it easier to see and touch and imagine the realities of deep time, geological forces, and biological evolution. Canyon rocks hold ancient days stacked atop days, years tree-ringed atop years, adding up to hundreds of millions of years. The canyon displays geological forces nakedly and thoroughly, the ancient mountains that melted into seas and were uplifted into new lands and into new rivers that began the land's return to the sea. The canyon is an archive of biological evolution, its limestone cliffs built mainly by life, by creatures turning water and sunlight into their shells and skeletons and flesh, into shapes that continued changing from era to era, growing more elaborate and capable and aware, growing into creatures who could look back into time and envision life's emergence, who could see the entire canyon as their Sipapu.

With the Sipapu and Ribbon Falls, with Vishnu Temple and Venus Temple and the Tower of Ra, with the visions of Christians and the studies of geologists, the Grand Canyon offered rich raw materials for the need of humans to understand their origins and meaning.

In every time and place and culture, humans have created creation stories. Perhaps it begins with the basic animal mind. All animals need to make sense of their environment and their relationship to it, need to create stories, mental programing about food and danger, about the land and weather and seasons and other animals, stories that store patterns of events and the varying outcomes of their own behaviors, stories that tell them how to feel and act. Perhaps some animals wonder where the world came from and why it is the way it is—we can only wonder, as perhaps only humans can wonder. Humans can see beyond surface events, see deeper patterns, and ask where those patterns came from and why they are the way they are, including why some patterns are generous or cruel. The world holds powerful forces that seem to deem it important that humans exist, that give us sunlight and rain and food and birth. Humans hold forces that urge us to go on living in spite of struggle and suffering and death. If humans sought answers to these mysteries by asking the rain and stars, they did not answer, and if the eagles and coyotes knew the answers, they were not telling.

Everywhere, humans have devoted a great deal of energy to creation stories. Creation stories are not just cosmologies to explain the workings of nature; they define human identity, make sense of our lives, and tell us how to live. They tell us why the universe exists and how we fit into it, why secret powers created humans, what we owe them and what they owe us. Creation stories are the foundations of religions, assigning the gods their roles and purposes. Creation stories define our relationship with the rest of creation, with the past and the future, with the sky and land and water, with plants and animals, with other people and groups of people, with birth and death, with good and evil. They ground human values and behaviors and hopes in the basic structure and purpose of the universe. They tell us why there's order and disorder and what we can do about it, or why we can't. They tell us why we should go on living.

Creation stories have been strongly influenced by the diverse habitats and lifeways of humans, growing out of mountains or deserts or prairies, growing out of hunting or farming, growing out of villages or empires. Creation stories propose every kind of god: animals and elemental forces and invisible mysteries, gods wise and foolish, omnipotent and limited, immortal and imperiled, generous and jealous, good and evil. Creation stories portray humans emerging from the earth, from the ocean, from a cosmic egg, from the body or mind of a god, or from nothing.

Humans have acted out their creation stories in thousands of rituals that align humans with the purposes of the cosmos and align the cosmos with human purposes. Failure to honor the cosmic order brings natural disasters, crop failures, social disorder, enemy attacks, disease, and death.

Humans have devoted enormous energies to embodying their creation stories in stone, in altars and temples that give ideas visible forms and inspirational power. Out of the ground, out of shapeless rock that the real creation story had hidden underground, out of the human need for order, humans quarried rock and hauled it dozens of miles, lifted and stacked rock into shapes that fit human bodies and emotions. They carved the rock into the bodies and faces of gods, into eyes that acknowledged human aspirations, into hands that could help humans. They painted the rock into elaborate murals of pantheons and theologies. Echoing music, the rock sang affirmations. Echoing priests, the rock spoke answers to ethereal questions. Into rock faces, human gazed lovingly and fearfully, trusting them with their deepest secrets and hopes, praying to them, lighting candles and incense before them, offering them gifts, killing animals for them, sacrificing humans for them, marching armies against distant temples with contradictory gods. Yet to the rock faces all the worship remained unseen and unheard. Humans were praying to seashells that had lived modest lives on the seabed eons ago.

If the greatest tribute humans could offer to their creation stories was to turn them into stone, cavernous spaces, stained-glass colors, and god-voice echoes, then consider the Grand Canyon's far grander stone and spaces and colors and river voice. The Egyptian pyramids would be dwarfed by the canyon butte called Cheops Pyramid. The canyon is a natural temple to the scientific creation story. Its river-fingerprinted rock faces are the faces of the true creator gods. The canyon's geological strata and evolutionary fossils played an important role in disclosing the scientific creation story, and today the canyon stands ready, like religious temples, to make our creation story come to life, to make time and Earth's power and life's creativity more real and awe-inspiring.

One person awed by the canyon's mythic power was astronomer George Ritchey, who, starting in the 1890s, designed and helped build some of the world's greatest telescopes, including the Mount Wilson telescope with which Edwin Hubble proved that the universe holds numerous galaxies and is expanding. Ritchey became enthralled by the Grand Canyon. In the 1920s Ritchey designed the most ambitious telescope ever imagined, and he insisted it had to be located on the rim of the Grand Canyon, at Desert View, because he believed this was the best place in the world for astronomy. His observatory would be a tower some twenty-five stories tall, with Greco-Roman architecture and statues, perhaps statues of Apollo, Venus, Jupiter, and the other Greco-Roman gods whose temples were directly across the canyon from Desert View. Ritchey must have been not just awestruck by the canyon but blinded by it, for in truth the canyon was a terrible location for astronomy, welling up

PROLOGUE

heat that would blur astronomical viewing. Edwin Hubble himself, interrupting writing his proof that the universe is expanding, came to Desert View to test it out.

The astronomers atop Ritchey's observatory tower would have been seeing in the sky the same names they were seeing in the canyon, in Venus Temple, Jupiter Temple, and constellations. Humans have filled the sky with thousands of gods and stories and built some of their greatest temples to honor sky gods, carefully aligning temples with the patterns of the sky so that humans could align themselves with cosmic order. The night sky inspires the same awe as the Grand Canyon, but deeper. The night sky is a canyon far deeper than the canyon, as deep as the mystery of human existence.

The astronomers atop Ritchey's tower would have been seeing in the night sky the same eons they were seeing in the canyon. The sky too holds strata, layer atop layer, era beyond era, event before event, all of it fossilized, a sky full of memories, adding up to a great story of power and change. They would see a supernova that happened at the same moment centuries ago that a cliff failed and crashed down. They would see rivers of starlight that had been flowing for as long as the Colorado River. They would see spiral galaxies as distant in time—275 million years—as the spiral fossils in the canyon's top stratum, the Kaibab Limestone. They would see galactic redshifts that revealed a dynamic cosmos as vividly as the Redwall Limestone. The canyon's deepest and oldest stratum, the Vishnu Schist, represents about one-seventh of the age and story of the universe. The canyon is a continuation of the story in the sky, the same story of deep time, deep forces, and deep creativity, a story that grew from energy into substance, from nebulae into rock and water, from fusion into metabolism, from obliviousness into wonder. Through humans the universe's story turned into creation stories. Temples and observatories sprang from the same yearning that created creation stories, a yearning that ultimately was the universe now empowered to question itself and seek answers.

Over the last few centuries the telescopes and microscopes and imagination of science have revealed a creation story far more elaborate than any story humans have imagined before—and a creation story that is not imaginary. The patterns in science's sky are far more sophisticated than anything any Stonehenge could detect—vast geometries of gravity, precise mandala symmetries of physics, family trees of stars. Its stars are not the eyes of lusty gods but the "lust" of matter to unfold the order seeded in its core. Real rivers contain more power and creativity than all the water spirits humans implanted in rivers. With star-born elements and nebular wanderings and earth wombs, life is more deeply born from the cosmos than either Father Sky or Mother Earth could offer. Science's creation story holds far more time than almost all religions had imagined, and holds a far larger cosmos filled with far more things and forces and activities, with layers of order upon order, with matter and energies strange and talented and powerful and deeply creative.

The new creation story is manifest in the Grand Canyon, vividly, better than anywhere. Time is here, the deepest time, time turned as solid as rock, unmistakably real. Space is here, the incomprehensible spaces that wrung energy into rocks and into this symbol of itself in which humans can feel appropriately dwarfed and awed. The stars are here, and not just because the canyon's high desert skies offer the clearest night skies people have ever seen, but because the stars forged the atoms of the cliffs aglow with sunlight. The elements are here, the same air and water and earth and fire that were flowing across Earth two billion years ago and that now have paused to portray a canyon before returning back into sand and rain. Life is here, for the canyon's sedimentary rock layers begin with the Cambrian era in which life exploded into millions of forms; life tells not just its past stories but the stories now growing from the rocks, walking on the rocks, flying over the rocks.

It's all here, all the ingredients of human identity, our deepest identity, our nature-defined identity, not the masks we wear in human society. Here we might be able to figure out the human meanings of the new creation story. The canyon is a theater in which you can test and act unfamiliar roles. Philosophical questions can turn into physical quests that may lead to bedrock. You can enter and interact with the eons, measuring them with your feet and kayak paddle, your heartbeat and tired muscles. You can feel correctly tiny and yet also begin to feel all the time and grandeur within yourself. You can see fossils and begin to recognize them within your own body. You can let the sunsets begin to recolor your own image. You can listen to the river and the wind until you begin to hear them whispering your ancient name.

Yet the first generations to look at the new creation story saw mainly what it had lost. They were living in religious creation stories, in a cosmos created intentionally, for the sake of humans, ruled by benevolence and protection. In the scientific creation story they saw an unrecognizable, forbidding, lonely place. Humans have always been lonely in the universe, even after we created gods to keep us company, and the emptiness of the astronomical universe imposed a much vaster loneliness. The new universe was also vast with time, which seemed far more preoccupied with rocks and dinosaurs than with humans. Instead of intention, the universe was run by the tension of physical forces. Instead of protecting life, the universe was so devoted to death that it built mile-high fossil graveyards.

The new creation story does indeed bring some losses. Yet the new cosmos offers its own strengths. First of all, it is real. After wandering in a maze of confusion and wishes for most of their history, humans have finally made contact with the real cosmos that created us. This cosmos offers a creation story richer, in its own ways, than all previous creation stories, full of ability, harmony, intricacy, and creativity. The power of the sky gods has been replaced by the vastly greater power of the Big Bang and the stars. The nurturance of Mother Earth has been reborn as the nurturing order of atoms, planetary orbits, geological moldings, and biological fecundity. Perhaps the longing and reverence humans applied to the sky gods and Mother Earth can now be transferred to the real sky and Earth. Exploring the new creation story and the place of humans in it is the primary cultural task of our time. We need to turn ideas into identity, to incorporate the new cosmos into our personal sense of reality, to go from naming the cosmos to letting it name us, to begin to feel it and live it, to discover the sky and Earth within ourselves, to turn it into art and celebration.

The Grand Canyon is a deep and abundant quarry in which to

find and construct human identity. Pursuing this goal honestly does require us to consider some of the losses involved. Can we fill an existential void with a geological void? The real human lives that go on in the canyon, including the struggles and despair and suicides, may help us assess the fit between the human mind and the new cosmos. In the canyon we can live the new creation story to discover if it is livable; we can mean it to find out what it means. If we listen carefully, the canyon may begin to tell us that its eons and depths and beauty are also our own. And perhaps it takes a surrealistic canyon to startle people out of feeling that there is anything ordinary about the world, startle us into recognizing that the universe is very strange, and so is our presence in it.

Because evolution saw no need to implant in humans a motion detector that tells us the ground is moving and not the sun, I had to prompt my imagination to see the genuine cosmos at work as sunset reached its peak brilliance and began to fade. The cosmos had continued painting canyon sunsets all along, even when humans saw only the sun moving and the doings of gods and dragons. I watched shadows emerging from their hiding places and growing and engulfing the cliffs. I watched my own shadows displaying unfamiliar shapes and merging into larger shadows. The canyon became one giant shadow and began merging with the darkness of the sky. The canyon was not being hidden by the night, but revealed. Night is the true face of the cosmos, and it was revealing the greater realities to which Earth belongs: the circlings of the solar system, the generosity of the stars, the grandeur of the galaxy, the energies born from the Big Bang. Rocks and trees and animals too merged with the night, revealing the identities they'd had all along.

I was sitting where George Ritchey's observatory was never built. But soon after he abandoned his plans, architect Mary Colter built her own observatory here, the Desert View Watchtower. She modeled it after Ancestral Puebloan towers that served astronomical and ceremonial purposes, and she asked Hopi artist Fred Kabotie to fill it with astronomical motifs, many copied from ancient rock art. The Watchtower was Colter's tribute to the Hopi people and their spiritual bond with the Grand Canyon, their seeing the canyon as the origin of humans. Kabotie's central Watchtower artwork is the story of Tiyo, a Hopi youth who rode a log down the Colorado River through the Grand Canyon and found the secrets of living. The Hopi creation story, through which humans sought greater order in the cosmos, was one more flowering of the order-pregnant, order-seeking forces that sprang from the Big Bang.

The stars came out, though they had been here all along, merely hidden from human senses, in the same way that humans lack the motion detectors to appreciate the gift of existence. Sitting on the canyon's jigsaw-puzzle rim, gazing into the canyon's mystery, I watched the stars emerging, becoming rivers of stars, canyons of stars.

The River

JOURNEYING ON THE WATERS OF CREATION

From a smooth, scientifically designed highway, we turned onto a road of dirt and rock and the sacred.

We were heading for Lees Ferry, our vehicles loaded with rafts and supplies for my long-awaited, private-permit, two-week river trip through the Grand Canyon. We were a science-minded group, with professional biologists and National Park Service interpreters, yet no one questioned my suggestion that we take some time and miles to cross paths with an ancient blessing. We stopped near sandstone cliffs from which boulders had rolled down. Across the boulders were more than two thousand images, carved over centuries by Hopis making a pilgrimage to the bottom of the Grand Canyon. Each of these symbols contained tens of thousands of footsteps and a dangerous climb in and out of the canyon. Like the rocks of Stonehenge or Easter Island, these boulders spoke of creatures hungry to find meaning. The sand grains chipped loose from these boulders had rejoined the endless cycles of sand that had buried them here long ago, while the sandstone that remained had entered a new world of yearning symbolism.

The Hopis' pilgrimage goal was the Sipapu, and, a few miles beyond that, salt deposits from which they would bring home salt. The Hopis could have found salt deposits closer to home, and much better grades of salt, but the goal wasn't just salt but salt from the bottom of the Grand Canyon, from the place of emergence. This salt was a spiritual nutrient, used in ceremonies. The pilgrimage was also a homecoming, connecting Hopis with their ancestors who, a thousand years ago, lived and died inside the canyon. The canyon pilgrimage is full of shrines, rituals, and secrets. I had a Hopi neighbor who went on the pilgrimage at age ten, guided by his uncle, and because he had grown up in a California city, the landscapes alone, the deserts and mesas and canyons, came as a revelation to him, and he finally understood how deeply his people and he belonged here.

We could not belong to this land so deeply, nor could we ask any gods for safe passage or supernatural rewards. Yet by detouring here we were confirming that our journey into the canyon would not be about mere recreation but also about creation, not about conquering nature but sharing the Hopis' respect, gratitude, and humility toward nature. We were locating ourselves on a map, a thousand years old, that showed the canyon to be a mythological landscape, with answers to ultimate questions about nature and human life.

When a drought forced Hopi ancestors to abandon the canyon and move to the mesas and springs where they live today, they took with them memories of not only the Sipapu and the salt deposits but the myth of Tiyo, a boy whose journey down the Colorado River earned the Hopis their most important bond with nature. The myth of Tiyo is the equal of the world's great myths of heroes making journeys, often journeys by boat, like that of Odysseus. It's possible the myth of Tiyo was inspired by real river adventures.

Tiyo lived near the canyon rim, where he often sat and watched the river flowing, flowing back toward the underworld from which it arose. The river's abundant water was a great contrast with the fluctuating springs and rains on which Tiyo's people relied and which sometimes failed, bringing much suffering. Tiyo resolved to follow the river to the underworld and meet the gods who controlled the world's waters.

Tiyo and his father cut down a cottonwood tree and hollowed it out. With offerings to give to the gods, Tiyo climbed into his log, and his father sealed it and shoved it into the river. Tiyo plunged through violent rapids and whirlpools, and after many days his boat came to rest on a beach and he stepped out.

Tiyo heard a voice calling to him from underground. Spider

Grandmother. The ground opened up and Tiyo presented his offerings to Spider Grandmother; she recognized his goodness and became his guide for his further journey through the underworld, full of dangers. Tiyo found the vast waters into which the river emptied. Tiyo entered the kiva of the Snake people, who guarded the secret of rain. Tiyo met a Snake maiden, whom most humans would find repulsive, but he saw her inner worth, proposed marriage, and took her home to his village. When her Snake relatives arrived for the wedding, the humans were frightened, but Tiyo convinced his people to dance with the Snake people. At the end of the dance an abundant rain fell. Ever since, the Snake Dance, in which Hopis hold rattlesnakes, has been their most important ceremony.

The Snake Dance is not an exercise in mere magic, summoning rain, but a sophisticated philosophical statement about the nature of the universe and the role of humans in it. Farming in a desert gives humans an unusually strong sense of both the blessings and negatives of nature. More than most peoples, the Hopis appreciate the nurturing rains and soil and seeds and growth, but they don't forget that nature is also a dangerous rattlesnake that could turn against you and kill you. Tiyo's river journey was dangerous but gave life. His Snake bride could manifest herself as beautiful or hideous. In Fred Kabotie's mural in the Desert View Watchtower, Tiyo's wedding ceremony is ringed by four snakes, three of them rattlesnakes.

The Colorado River twists like a snake. Even rafters with no mythological interests feel they are embarking on an epic journey. It raises questions of life and death. It can question human identity—some adventurers emerge as different people. The canyon remains powerful, even if it speaks to us in a different language than it does to the Hopis.

We could still speak to the river and the canyon: *May we journey like Tiyo, brave yet humble, watching and listening, into the depths of the earth, into the heart of creation, into time and the elements in their own home, into the source of ourselves.*

A shadow was rushing toward me, an absence signifying a presence, an eclipse whose sudden approach tricked from my mind a recognition:

Raven! whose certainty was greater than I felt when wondering what my own shadow signified.

On the water the raven appeared twice, as a shadow and as a reflection, yet the water, unlike the human mind, which cascaded with images to connect the raven with a memory, a meaning, or at least a name, made no effort to interpret the raven and left it a simple presence, another flowing.

Of course, the raven was also a mind working to understand what it was seeing. From its ledge high on the cliff, the raven spotted a few humans moving upstream, sometimes wading in the water, sometimes legging through the shore brush and boulders and mud. When the humans were in the water their legs blurred and warped and flowed like water, and then they emerged and stopped wavering and became solid again, emerged into the identities humans imagined to be real and reliable.

The raven could never figure out why the humans came here. They did not come very often. They always stopped at the same place, a dome of rock, hollow, with water inside, water bubbling up and flowing out. The raven had tasted this water and it was terrible. Sometimes the humans carried feathers and left them inside the dome. After the humans left, the raven landed atop the dome and peered inside and studied the feathers. They were never raven feathers but often monster feathers, monster eagles. The raven found all this very strange.

We had walked several miles from the Colorado River, up the shallow Little Colorado River, whose milky, light turquoise color felt surreal. The Sipapu, too, was surrealistic, an odd dome maybe twenty-five feet tall, yellow and brown and white, with golden water trickling out. In the top was an opening.

This was the place of emergence, where humans had first emerged into this world, the fourth world life had inhabited. The first three worlds were underground and dark and full of strife. Tawa the sun god had intended people to live harmoniously. Out of formlessness he had gathered the elements and mixed his own essence into them, creating the world and life. Yet the first life was insectlike and did not understand their origin and purpose and fought among themselves.

Tawa turned them into animals, but they, too, quarreled, so Tawa turned them into humans, who should be able to understand his good intentions. When the humans, too, lived in strife, Tawa sent Spider Grandmother to lead good-hearted people up a reed through a hole in the sky, the Sipapu, to a world of light.

I looked around, as if I was one of the first humans to emerge and see this new world of strange shapes and colors. If I wanted to, I could have quickly located myself on overlapping maps of familiar realities, of geographies full of miles and bends, of sciences full of geological strata and calcium carbonate springs, of histories full of peoples and cultures and mythologies. Yet for the moment I wanted to linger in the unfamiliar, to see it with astonished eyes, to see not concepts but naked presence, for this vision of strangeness was our truest recognition of the world and of ourselves.

Just then a shadow swooped over me and I looked up, and for a moment my eyes and raven eyes looked right into each other and we recognized, across the boundaries of lives and habitats, across the differences of arms and wings, the need of animals to find significance.

As I stepped away from the Sipapu and back into the Little Colorado River, I noticed the Sipapu's golden water, its amniotic fluid, flowing into the river's blue-green water, dissolving into tendrils and swirls and vaguer colors, merging into a greater identity. The Little Colorado would never be the same. Human eyes could not see it, but a chemical assay would show, even four miles downstream, that the river still held Sipapu water and thus a slightly different quality. And what would a mythological assay reveal? The Little Colorado now flowed with Earth's secrets, with the powers of creation.

As I walked in the river, up to my knees, I felt Sipapu water. It was making my limbs strange to me, heavy and slow and unbalanced, as if trying to remind me that a human body is always a strange thing. The water mirror showed me my own image, constantly rippling— the Sipapu waters said that all along I had been composed of Sipapu ripples. I was being baptized in the river of creation.

In the water and on shore were cobbles that had fallen from the cliffs as jagged chunks and been polished smooth and pretty by Sipapu waters, stone eggs arranged by size and shape into mosaics.

Growing from the cobbles were plants that Sipapu waters had polished from within, into shapes more elaborate than cobbles. In the river, Sipapu-polished fish darted on inner currents that steered them to uphold order.

When I got back to the Colorado River I watched the blue-green waters of the Little Colorado beginning to mix with it, a mixing that would take miles to complete and that would change the quality of the river for the rest of its way through the canyon. The rest of our journey would be conducted upon Sipapu water. Its power would bob our boats up and down and steer them back and forth and along the pathway it had helped carve. It would give us miles of calm beauty and then turn violent and try to drown us. With huge waves it would immerse us in both creation and chaos. It would grace us with sunset and sunrise reflections. It would sparkle with stars all night, stars flowing from their greater Sipapu. It would mirror our own faces, faces wavering, ghostly, deeply true. It would offer us, whether we noticed or not, ceremonies to initiate us into its secrets.

Two miles downstream from the river confluence, the Colorado River enters a stratum of dark red sandstone, formed more than a billion years ago at the changing boundary between the sea and land. As the sea advanced onto land it carried lots of salt, and when the sea retreated and its abandoned lakes dried up, it left lots of salt. Over millions of years the salt accumulated, then got buried and remained buried while geological life and then biological life went on above it. One day the salt's coffin lid was opened and it felt the sunlight it had not felt in more than a billion years.

From an overhanging cliff near the river, the salt emerged and formed stalactites and stalagmites of salt. Occasionally a butterfly landed on the salt and licked it and said that it was good. Birds tasted the salt and said that it was good. Mice and lizards tasted the salt and said that it was good. One day a human gazed at the salt, tasted it, and said that it was sacred.

As the salt flowed into human bodies it looked for an explanation of why it had become sacred. It found a sea a lot like the sea it had belonged to long ago, a salty sea. The sea had wrapped itself up and flowed onto land, flowed into new and strange currents, yet remained in love with its ancient salty identity. Now the salt flowed throughout human bodies and was inducted into the secret societies of life. It was stationed along cell membranes with just the right assignments of sodium and potassium molecules to set up an electrical gradient that allowed cells to function, muscles to contract, hearts to beat, and neurons to fire. It was carefully applied to set the chemical formula of blood, the pressure of blood flow, the flow and potency of hormones, the performance of organs, and the sensitivity of senses. When the salt permeated the brain, it found and reinforced a dance of images and memories and values and desires. It found questions about this living body, this puddle, where it came from and why it felt itself to be exiled from something larger. It found judgments that this puddle was a good thing and should be continued. It found rituals for helping life continue, rituals in which salt was an important symbol.

In Hopi pilgrim mouths, the salt was eaten. From the canyon, the salt emerged. With its new legs and bodies and minds, the salt climbed step by step, rock by ledge by cliff, tilt by balance, discovering how hard it was to be a living body. The salt got tired and pumped more salt into muscles but got tired again. The salt sweated onto foreheads and arms. The salt craved salt and opened its pouches and ate more salt to support not just biological shapes but mythological shapes, to crystalize answers to ultimate questions. From a billion years of innocence and rest, the salt emerged.

(Out of respect to the Hopis, the salt mine is off limits to others, as the Sipapu would be a few years after our visit.)

EMERGENCE

Some of our journey in this book will be by river, and some by rock and trail.

Reaching out, reaching into space and time, I touched a galaxy, a sparkle in the blackness.

I was touching the Vishnu Schist, the rock layer at the bottom of the Grand Canyon and much of the continent. I was touching rock 1.8 billion years old, about one-seventh as old as the universe. Ordinarily, such time and space lie beyond the reach of human senses and understanding. Our eyes can barely see the nearest galaxy, Andromeda, at only 1/700th of the distance to 1.8 billion light-years. Yet here, in this rock, deep time and space had crystalized into solid form, allowing me to get a feel for them. The Vishnu Schist was night-sky black and sparkling, flecks of mica remembering the forces that had transformed mud and sand and clay into rock. I pressed my fingers against the schist and felt the reality and strength of time and creation.

Vishnu was an appropriate name for the rock that upholds all the rocks of the canyon and continent, for Vishnu is the Hindu god who upholds the universe. When demons and forces of chaos threaten cosmic order, Vishnu emerges, in various avatars, and fights them off, restoring order. Vishnu permeates the Ganges River, turning sacredness into a physical force people can embrace.

Somewhere inside the Vishnu Schist, but effaced by the tectonic pressures that formed it, were elements that had been alive, basic cells, halfway through life's procession on Earth, on its way to becoming beings who would turn a river into a god. Those cells were not so far away, for they lived on inside me, they continued living their own ancient lives, loaning their learning and abilities to me for a moment.

Turning my attention from my finger to my feet, I felt the Vishnu Schist beneath me, upholding me. I started up the trail. With every footstep I was testing and confirming the strength of the schist and the pink-white granite snaking through it. Like the granite, the trail snaked through the schist, seeking out its weakest seams, its most human-friendly routes. My course was being steered by long-forgotten geological events, by sand and mud piling up for eons, by continents colliding, by land rising and fracturing rocks. I might imagine myself to be in charge of my own motions, choosing every step, but no, I was another grain of sand being broomed along by ghost continents, ghost magmas, ghost rivers, ghost winds. Those forces had also steered life's evolution, swerving it from shape into unpredicted shape, now into this shape I imagined to be my own.

At the top of the schist, about a thousand feet above the river, I took one footstep and time-traveled across 1.2 billion years. The rock layer atop the schist was suddenly much younger, "only" 525 million years old. This gap had once held thousands of feet of rock layers, but they had eroded away, raindrop by raindrop, grain by grain, day by century. In this gap continents formed and broke apart, Himalayas rose and disappeared, and numerous great rivers carved great canyons. Placing my hand on the gap, partly on the schist and partly on the layer above, I encompassed one-quarter of the age of Earth and one-eleventh of the age of the universe. Time had also erased those memories of itself from human minds, though we still contained all those eras and all that work. Canyon hikers are often amnesia victims, walking through our own house with no recognition of it or of ourselves.

Now I was walking on sandstone, and imaginatively wading along an ancient seashore with rivers pouring sand off the continent and piling it into beaches and shoals and the seabed, with currents and tides and storms rolling sand up and down, sorting it by size, layering it. Salts that had flowed in that sea were now leaking out of my face.

When I crossed the boundary between the Vishnu Schist and the sandstone, I was crossing from a vertical world into a horizontal world, crossing out of the schist's compression lines and granite intrusions and into the sedimentary layers above, layers laid down line atop line, tiny lines adding up into huge colored bands, which helped a human mind better imagine the passage of time.

I passed from the tan sandstone into green-black shale, from cliffs into slopes, from strong rock into brittle, cracked, shambled rock. This shale had been laid down farther out at sea, farther from the mouths of rivers, beyond the range where rivers could propel sand, in the range where only lighter sediments could drift. I was walking through mud about 515 million years old. Mud did not make the strongest rocks, and when this shale was exposed to erosion it disintegrated faster than other layers, bringing down the cliffs above it, opening a much wider canyon.

I passed from the shale to a grayer rock, a limestone. It had formed even farther out at sea, beyond where all the land sediments had settled out, where the only sediments left were sea sediments, the

CANYON AND COSMOS

shells and bits and juices of sea creatures, rich in calcium and carbon. Tiny shells had added up to a limestone shell hundreds of feet thick, the shell of turtle Earth. The limestone was still imprinted with the shapes and trackways of some of these creatures. I felt those trackways heading toward me, felt those shapes inside me, and I imprinted on their dust the latest evolutionary news.

I stepped onto another limestone, with a color and texture and strength different from the first limestone. The trail began climbing steeply and switchbacking more, for this limestone, formed in the deepest sea, formed the tallest cliffs in the canyon. I was deep underwater. I felt deep currents flowing, flowing for eons, still flowing today, for that water now flowed inside me, confined by stranger shores. That water had flowed into many new patterns inside life, pulsing and growing and swimming and swarming, tossing a tsunami of life onto land, where flowing was not so easy. Now my water had to flow uphill. Climbing through these cliffs reminded me that life had always worked hard against entropy, every day and in every life, and in climbing the evolutionary trail. I felt the enormous mass of living and dying that had built this rock, and built me. I contained multitudes.

The trail's steepness eased, and it grew redder and more soil-like. As I hiked, my boots continued changing color, being painted by the geological eras. For the next thousand feet of elevation I walked through a coastal plain, with the ocean coming and going, leaving a mixture of marine fossils and reptile tracks.

The trail and cliffs turned even redder. I was far inland, where rivers and streams dropped silt onto swamps and floodplains full of ferns and ginkgo trees. Insects swarmed and buzzed. Amphibians scurried out of the way of reptiles, which were growing larger. Life was also moving toward mammals, toward myself.

The cliffs turned tan and the trail turned sandy, and I was far away from the ocean, in the middle of a desert hundreds of miles wide and hundreds of feet deep. Hot winds blew here for millions of years, driving sand dunes onward, piling them into crests and crescents and blowing them down, moving them along, piling them up again. In the rocks I saw stripes sloping this way and that, clusters of lines

26

thickening or thinning, petrified sand dunes that told me which way and how strongly the wind was blowing 280 million years ago. Reptile footprints and tail streaks told me of how one creature felt cold one morning and climbed to the top of a sand dune to warm itself with the same sun I was feeling. Now this sand dune was unraveling back into sand and feeling familiar winds and forming tiny dunes, which my boots blew in new ways.

The trail changed color and texture and gradient again, and I was wading through water. The ocean was returning, deepening, and now it was filled with life, with crinoids, sponges, clams, snails, nautiloids, trilobites, fish, sharks, corals, and numerous plants. In the cliffs I was seeing lots of fossils, life holding steady, life always changing.

As I climbed and saw the rock layers adding up, I started seeing time more clearly, it turning from an intellectual abstraction into something as real as rock. Seeing myself so small against the cliffs, I felt the massiveness of time. Seeing sand grains and seashells piled up one by one, I saw the smallness of time, how time was not just a force of eons but of moments and tiny events. I saw time being counted by sand grains, by raindrops, by gusts of wind, by ocean tides, by rocks cracking apart, by seeds opening, by plants growing, by dragonflies buzzing, by new mountains and new forms of life arising very slowly. I began to see, actually feel, all of that time within myself. Climbing the rungs of time set time ringing in my heartbeat and muscles. The eons that built these cliffs had also built me and still resided in me and now the eons were counting time with my footsteps. Time had become biologized, time lived and time remembered, time regretting the past and fearing the future, endless time yearning for more time. Time had loaned me a raindrop of itself to call my own. Yet in me unstoppable time was getting a bit tired and wanting to stop.

Step by step, rock, too, was becoming more real, hammering its strength into me. I also saw rocks eroding everywhere, and my boots were contributing to it, freeing sand from half a billion years of confinement. Stratum by stratum the powers of creation and destruction spoke more clearly. I felt tectonic forces raising the land out of the sea and creating mountains and volcanoes, and saw rivers carrying mountains and volcanoes back to the sea. I saw the sea rising into clouds and speaking as wind and lightning and raining down and rushing back to its true identity. I saw the cycles of particles becoming rocks and returning to particles and becoming new rocks. I saw the cycles of the earth, the ever-changing landscapes. I saw life doing its best to live and yet turning into rock, then getting free and joining life again. Step upon step, I was seeing stratum upon stratum within myself. I felt tectonic forces raising me, oceans and rivers flowing through my veins, and mountains sifting themselves into my flesh. My calcium and carbon recognized that they had once flowed through those seas, and had even eroded from these very cliffs a million years ago. The canyon was walking. I felt the rivers and winds that had flowed across these ancient lands now pushing me up the trail. I felt the erosion that carves out canyons now gnawing away on my energy and body.

Step by step, stratum by stratum, I was seeing geological forces transforming into biological forces, seeing life transforming from Precambrian cells into greater complexity and abundance. In the Vishnu Schist life was invisible, requiring a chemical assay to find any trace of it, but the canyon's top layer, the Kaibab Limestone, was bulging with fossils full of artistry. All of that evolution inhabited me. Cells that had taken two billion years to figure out how to combine into bodies were holding me together—I hoped. Seashell calcium and curves strengthened and graced my bones. Half a billion years of antigravity and balancing strategies, refined and passed from species to species, were gyrating in me. I was a puppet whose muscles were being pulled by trillions of strangers. My temperature was being regulated by tricks that had been passed from the oceans into forests and into deserts and up to mountaintops. Fish hearts were making my heart beat; fish fins were wiggling my fingers and legs. My lungs remembered being gills trying to decide if they should leave the sea for good—and at this moment they weren't sure if they'd chosen the easiest course. Through ears that had listened to Jurassic winds and pterodactyls, I was hearing ravens calling. The trilobites who had invented eyes were gazing through my eyes, amazed. Primates gripped my hiking poles. I felt vast crowds of animals within me, melded together, still trying to go somewhere. The same power

THE RIVER

that had moved reptiles up sand dunes 275 million years ago was driving me now, a power larger than all of us.

The canyon's cliffs and fossil record ran out before it got to the age of dinosaurs or true mammals, but that record was written within me. I was loaded with fossils; I was a miniature, animated Grand Canyon. In hiking the canyon you do wonder how you'll hold up. You know human bodies have limits and sometimes falter and fail here, even die. I thought about all the layers of order upholding me. I was feeling heavier than usual, but this included a solidity that ultimately rested on the solidity of atoms. Throughout me, atomic nuclei were holding tight and electrons were holding to them at highly precise and reliable speeds and orbits. I was feeling my atoms, with a strength forged in the Big Bang and stars and supernovae, defending their solidity, even as they linked up with fellow atoms to build larger solidities, including vast fabrics of molecules. Atoms were also the foundations of far larger circles around me, the orbits of the solar system and the galaxy. I was surrounded by a vast, intricate order, by motions within motions, by circles within spirals, by cosmic harmonies. In between cosmic harmonies and atomic harmonies, fed by both, was another vast harmony, the biological cosmos, the circles of cells, also full of motions, motions upon motions, motions building intricate order, order upon order. I felt this swirling cosmos within me, rushing to uphold and move me, even when I did not recognize it or direct it or thank it. I felt atomic harmonies and cosmic harmonies flowing into my cellular harmonies and merging, felt the energy of the sun jumping from atom to atom and propelling my heartbeats and motions and consciousness. I was the nexus of harmonies incredibly patterned and trustworthy and skilled and creative, harmonies upon harmonies, harmonies overflowing into greater harmonies, harmonies ultimately mysterious, with no good reason for existing, harmonies that had inexplicably decided that I, too, should exist.

One last step, and the cliffs ended. I was looking over flat lands. I was standing atop the final ocean intrusion onto this part of the continent, the final layer of marine fossils. Far above my head, in rock layers now gone, began the worlds of dinosaurs and mammals. I looked around and saw wildflowers, chipmunks, and songbirds, acting as if this was just another normal day on Earth, not seeing the enormous time, work, and lives that went into creating them.

I looked back into the canyon, at all its colored bands, merely pretty to the eyes but deeply powerful to the feet and mind. Hiking the canyon isn't about seeing yourself in the canyon, it's about seeing the canyon in yourself; it's not about feeling strong and accomplished, it's about feeling the strength of the forces that accomplished you. I looked back not at a mere trail but at my own trail, at footprints that had continued changing shape until they became my own, at a canyon of emergence my hike had recapitulated, a canyon from which I had emerged as a deeply layered life.

The Emergence of Spirals

Instantly, when I saw the overhanging cliff full of rock art, my mind jumped into making order, deciding: bighorn sheep, deer, human, cloud, lightning bolt, cornstalk, snake, spiral. I knew I was violating the caution of archaeologists about interpreting rock art, for a human-like figure could be some sort of spirit, and a wiggly line might not be a snake. But I couldn't help it: the human mind is eager to create order, to recognize snakes before we get too close to them.

The human need for order had created this rock art a thousand years ago. The artists put a lot of effort and skill into it, over many years. Some figures were painted with fingers dipped in iron pigment, pigment gathered far away and carried here and mixed with animal fat, figures that might hold faint remnants of fingerprints. Other figures were carefully hammered into the rock. I studied and admired this artistry, the lines and shapes, the organization, the colors, the feeling and imagination. I wondered what this art had meant to its makers, whether it was indeed art, or recordings of nature, or part of religious ceremonies.

I studied the spiral. Spirals are part of rock art all over the world. Sometimes the spirals have heads, suggesting coiled snakes. Then again, rock art in Ireland, a land with no snakes, is some of the most spiral-devoted rock art anywhere. The prominence of spirals across distant and diverse cultures seems to express the human mind's universal order-seeking tendencies. From rock art, spirals went on to

THE RIVER

star in architecture, mosaics, painting, pottery, and jewelry. Numerous religious shrines featured spirals—spirals suggest the idea of a journey. At Chaco Canyon, the religious center of the same Puebloan culture that lived at the Grand Canyon, astronomer-priests carved a spiral to mark the arrival of the solstice, to celebrate the order of the universe.

I also noticed a spiral that was not made by humans. This cliff was limestone rich with fossils, with the vague shapes of sponges and corals and seashells. Some of the rock art was painted or inscribed over the fossil shapes. The two blurred together into a mosaic more than a quarter of a billion years old, started by ocean hands, seashell fingers.

The fossil spirals held greater mathematical skill than the human-drawn spirals. They were nearly perfect logarithmic spirals, repeating the same dimensions over and over but with expanding size. The fossil spirals had been drawn by the same unconscious genius that draws spirals in pine cones, leaf arrangements, insect eyes, seahorse tails, and antelope horns, all of which were destinations of the spiral staircase of DNA. These fossil seashells had materialized out of invisibility, out of the calcium carbonate particles diffused through the ocean, which life arranged into lines, circles, and more complicated shapes, shutting out shapelessness, coloring them in. Wavering in this ocean were the shapes of future life. I saw these fossil spirals creeping along the ocean floor, ever questing. I saw their beauty, yet could not forget these were helmets to defend soft lives from being devoured. These defenses were so strong they had outlasted an ocean, easily outlasted the humans who had drawn rock art spirals.

This rock art flowed out of a creativity deeper than biological creativity. This rock art rested upon the art of rocks. For four and a half billion years rock had never stopped remolding itself, filling itself with lines and curves, speckles and sparkles and colors. Rock treated life as a new raw material for building rock. Rock created beauty that life would one day quarry into statues and cathedrals. Rock offered not just the stage upon which life acted but the off-duty rock that flowed into life and became its bodies, became hands grinding iron into paint and adding to rock a strange new sedimentation not from gravity and not from dying but from life yearning to settle questions.

This art flowed out of a fire even deeper than magma. It embodied the ordering tendencies in the heart of matter, the forces that shaped particles into atoms and shaped atoms into galaxies and stars and rock. These creative forces announced their presence and power in the first instant of the Big Bang and have continued unfolding new forms and new forces, unique planets and trillions of species, with every individual life unique, every day generating new thoughts and feelings. The Big Bang grew hands that could portray the Big Bang energy now in lightning bolts and animals running. Every speck and line of this rock art had been part of the Big Bang and contained the entire history of the universe, its endless searching and finding and building. This was the fire universe, the rock universe, the coral universe, the blind universe painting its self-portrait.

Yet from the birth of the universe and the first unfoldings of order, there was opposition, confusion, waste, and failure. There was energy that exhausted itself, motion that went nowhere, particles that never connected, formations that stopped short, possibilities that never happened. There were atoms that never joined galaxies, gas clouds that never became stars, stars that never birthed planets, planets that never held oceans, oceans that couldn't create life, and life that never got beyond the simplest cells. There were forces, events, and probabilities that actively worked against order: stars that sabotaged solar systems, ice ages that never ended, and mass extinctions that steered life away from greater forms and intelligence. Every life risked deformation, predation, disease, and accidents, and was guaranteed to die anyway.

In spite of all this instability and opposition, the universe's push toward order was so powerful it could not be denied, not everywhere, not forever, and indeed it overwhelmed chaos and converted chaos and built layer upon layer of order, sometimes in a rush, as if it knew all along what it was meant to do, sometimes after long experimentation. The emptiness erupted with shapes.

In the rock art spiral I saw the outrush of the Big Bang summoning itself into billions of spiral galaxies. I saw spiral galaxies reproducing themselves as nebulae spiraling into stars and planets, into hurricanes and river whirlpools. I saw oceans spiraling themselves into seashells and rolling through millions of generations. I heard the cochlea of the human ear translating the soundwaves of tornadoes into terror, translating the soundwaves of scroll-tipped violins into joy. I looked at the whorls of my fingerprints and saw I was the child of galaxies and seashells, I was their hands for creating rock art spirals, Chaco Canyon solstice markers, and Van Gogh starry nights.

Humans need symbols, we think and feel through symbols, and we have always wanted symbols of the cosmos and our place in it. For the scientific cosmos, for the Big Bang turning into galaxies and seashells and violins, a perfect symbol is the spiral, which begins at a point, unfolds steadily, makes a journey, and forms graceful patterns, still progressing, yet ultimately remains a question mark.

I felt within me a giant spiral, glowing and spinning, the Milky Way galaxy, its spiral arms become my arms, its motion moving me, its stars burning in my flesh and mind and gaze. I saw our galaxy as a giant question mark that had embedded itself in humans and taught us to see mystery and to ask questions and never be satisfied with our answers. I felt in our hearts a giant black hole, gnawing away. I heard the universe, after its long journey, arriving back where it began, in mystery, but now conscious of it.

As dawn spread down the cliffs, a rose of sound arose from the canyon bottom and rose toward the light, echoing off the cliffs. The sound came from a conch, a spiraled, foot-long, five-pound seashell. A woman lifted the shell with both hands, pursed her lips around the hole cut in the end, and blew into it, and its thick-walled chambers magnified her breath into a deep sound that reached far, even into sleep, announcing that breakfast was ready, greeting the day. Coming from a sinuous creature of water, the sound seemed right for a river journey.

Conch shells are widespread in the world's oceans, with many species, and humans have been enchanted by them for thousands of years. They show up in Mayan murals, in Japanese music, in religious

CANYON AND COSMOS

ceremonies, in weddings and funerals, and they got buried in the graves of kings. Buddhists carried conch shells to Tibet, carved symbols into them, and used them to summon people to prayer. Hindus placed a conch shell into the hands of Vishnu, the preserver. Rafters blowing conches amid the Vishnu Schist are hearing a deeper music than they know.

The air swirling out of the conch shell had roamed over the seas and continents for eons, formless and wild. It had blasted out of volcanoes, raged as storms, and pushed sand across deserts. Yet now this air was imprinted with order, with vibrancy, so different from the air around it. Its patterns, learned from one wave of a wave train of shells that had been rolling for half a billion years, flowed across the sounds of river and wind, informing them of form. It was a meaning amid incoherence, proclaiming not just conch but consciousness.

The conch call flowed up the cliffs and rippled slightly, spiraled slightly, on the fossils of the first seashells, its own infant footprints. It flowed higher and rippled off younger shells, the stone wedding rings of life's commitment to continuation. As the sound spread up the canyon, ravens and coyotes and bighorn sheep noticed it: the sound flowed into the conches in their ears, where the ancient sea and ancient life lived on, where the sound was greeted as a voice of life.

The conch call flowed skyward, diffusing, becoming inaudible, subtly reorganizing the sky. It headed toward space, trying to announce to the spiral galaxies that their formative genius had risen further into spiraled life-forms and into the confused yet still ascending voices of ascent. If humans raised conch shells not to their lips but to their ears, they would hear the pulsing ancient sea that already included their own pulse; they would hear the pulse of galaxies.

THE SHIMMERING OF CREATION

The cliff in front of my kayak was alive with energy, with lights dancing, lights constantly moving and flashing. With the angles right, with the sun behind me and a cliff in front of me and the river rippling modestly, the cliff had become an aurora borealis, pulsing entrancingly. The lights were ghostly, at least in the sense that Isaac Newton, enchanted by the powers of light, named its secret rainbow the "spectrum" from the Latin word for specter. The lights were ghostly in the sense that the world is full of strange energies. I float there, float upon strangeness.

Light is a central element in creation stories around the world. In polytheistic religions the sun god is often the creator god, who triumphs over a long reign of chaos and darkness and cold. Creation may never end, as every night and every winter the sun has to contend against the demons of darkness again. Sometimes the sun is the eye of god, honoring the human dependence on vision and light. For the Puebloans, creation consisted of emerging from a dark, lower world into the sunlit world. In monotheism a more omnipotent god creates a more reliable world with "Let there be light." In mystical traditions, divinity is spiritual light. Everywhere, the human soul is represented by images of light.

It was mythic, then, that after science had led humans steadily away from mythological creation stories, it arrived at a creation story that gave humans the greatest breakthrough of light they had ever imagined, and gave the stars far larger domains.

The light shimmering on the cliff was the light of the Big Bang, energy that had endured across 13.8 billion years and been transformed many times. It was light where there could have been only darkness forever, light that could have worn itself out long ago, light that could have lacked the power to create anything, light that might never have been seen by anyone. Eye floated there, I was x-rayed by light, diagnosed as vastly improbable.

At the Big Bang, the light of creation instantly revealed its commitment to creativity and patterns, including its own precise speed. Big Bang energy solidified into particles and larger particles and systems of particles. It revealed powers like magnetism and heat and gravity. It swirled into nebulae and galaxies and stars. With every star ignition, the Big Bang birthed its babies and rebirthed its light. It turned into bodies that light moved in ways light had never dreamed of moving. It turned into eyes through which the light of creation finally recognized itself.

I was there, all of me, part of the Big Bang, every energy of me, every particle, every future atom, every future cell, every future motion, every future glance and thought. I was there as a dream, a gleam in a quark's eye, a quantum karma roulette ball. I was conceived there and helped give birth to the universe. My particles contributed to the Big Bang's power and light and expansion and creativity. My particles were helpless in that chaos, swirling madly, yet helped shape it into order, laying the foundations of a cosmos. My particles still held that energy, locked deep in atomic memories on which the universe still draws and builds. My warm flesh was the afterglow of a hundred billion degrees. My particles were ancient, loaded with 13.8 billion years, yet still tireless, steady, pulsing, flying into the future. My particles were creator gods only pausing in me. My particles were ready to tell me epic stories, help me remember my own deeper story.

In the shimmering cliff I saw a mirror, saw my own reflection, not my merely human self but the Big Bang embodied. Somehow, the strange impulses in the Big Bang had become my impulse to journey down a river to read the journey of the universe.

RELATIVITY

Albert Einstein stood on a rock slab with a big crack in it and looked into the Grand Canyon, at the Colorado River twisting, at one of the biggest rapids. Einstein found the view a bit dizzying. He preferred to live in the universe of mathematical perfections. The rapid did seem to have some mathematical order, with regularly spaced waves steadily diminishing in size, yet it also appeared to hold too much chaos, the waves building to irregular shapes and breaking at irregular intervals. They were vastly inferior to electromagnetic waves, which had absolutely perfect shapes and speeds throughout the universe. And why did the river insist on curving back and forth, when it could save a lot of energy by going straight?

The main reason Einstein felt unsettled was that this rapid was dredging up messy feelings about his eldest son, Hans Albert. A decade before, Hans Albert had decided he was going to become an engineer, and Albert took this as a rebuke of his own values, of the nobility of pure science. Poor Hans Albert was only trying to avoid living in his father's godlike shadow. Hans Albert was reverting to his grandfather, who was in the electrical engineering business and

THE RIVER

who, when Albert decided to study theoretical physics, declared that theoretical physics was "philosophical nonsense" and demanded that Albert become an engineer. Hans Albert, after years of designing bridges and factories, had decided to become a hydrological engineer, studying and altering rivers. Albert felt that water was too messy and that hydrological theory was so primitive it could barely be called science.

Hans Albert had been fascinated by the messiness and power of water ever since, as a youth, he'd walked to school along the banks of the Moldau River and watched it surging through locks. Sometimes Hans Albert lingered at the river so long that he got home late and was scolded by his mother. Sometimes Hans Albert lingered at the river to avoid going home, where his parents were often quarreling. Hans Albert became so fascinated by the messiness of water that he wrote his doctoral thesis about it: "Application of Probability Theory to Sediment Transport."

To Albert, who had always tried to exclude from his life the "merely personal" and the messy human world, probabilities were never good enough, not in quantum mechanics, not in un-Bach music, not in God's mind, certainly not in his other son, Eduard, whose years of schizophrenic behavior would soon get him committed to a psychiatric hospital. Einstein's flight from the chaotic "merely personal" had generated his relativity theory, which insisted that the laws of physics never varied even if human viewpoints or motions varied a lot, even if time itself had to be sacrificed and became variable. The laws of physics were sacred.

Perhaps the Grand Canyon also made Einstein uneasy because it had started out small and grown bigger and bigger. Einstein had just come from Pasadena, where he had met with Edwin Hubble and reviewed Hubble's evidence that the universe is expanding. Einstein's own general theory of relativity had required the universe to be expanding or contracting, but Einstein had found this idea so bizarre and distasteful that he had invented a force that overruled it and left the universe stable and eternal. If only he had trusted his own equations and not his desire for stability, he could have predicted the expanding universe. Hubble's observations of redshifted

37

galaxies had "smashed my old construction like a hammer blow," Einstein publicly admitted, making newspaper headlines around the world. Now Einstein was staring into a canyon that seemed to personify the expanding universe, a canyon whose redshifted cliffs blazed with motion and change and instability, whose massive piles of rubble denied a steady state, whose river flashed mere probabilities at him as it rushed toward entropy. The night sky glowing over the canyon was no longer the peaceful and eternal sky preferred by poets and priests and astronomers alike, but a raging river transporting its sediment of galaxies.

So this is Time. Einstein saw time in the cliffs, actually felt time in the rock on which he stood, not theoretical time, not mathematical time, not relativistic time, but time solid and strong, resisting his gravity, upholding his life. Time was monumental. Maybe for the first time, he saw time as beautiful.

At that moment a newspaper photographer snapped a photo of Einstein and his (second) wife, Elsa, standing on the cracked slab, him on one side of the crack, her on the other. Seven decades later I became curious about where this photo was taken, and from its glimpse of the river I tracked it to Hopi Point and found the cracked slab exactly as it had been in the 1931 photo.

Perhaps the river Einstein had seen from here was the Hopis' river, Tiyo's river, an imperfect and dangerous river, meandering like a rattlesnake, not straight like light, a river Tiyo endured to reach the Snake people and learn the secrets of rain and growth and life. Einstein's biographers have suggested that his obsession with finding a perfect order behind nature's appearances came not just from his own personality but from his Jewish psyche, with its radical idea that the world was ruled not by numerous flawed, contradictory gods but by only one perfect god with a master design. Einstein often said that he wanted to know how God had created this world, though he had discarded the traditional God and made physics its substitute.

Hans Albert Einstein, too, would contemplate the Colorado River. When he'd started his career, hydrological theory was indeed primitive, using models of ideal fluids that couldn't come close to the enormous complexities of real rivers and their sediments. Hans Albert

THE RIVER

made major contributions to more sophisticated mathematical modeling, especially how different kinds of sediments were transported down rivers and settled out.

In the Colorado River Hans Albert saw enormous complexity. He saw a river full of sediments—some people called it dirt—of many kinds, behaving in many ways. In his father's cosmos, gravity was space-time warping with mathematical elegance, steering planets in steady curves. In the Colorado River, gravity was chaos, pulling water down, pushing water against water, making water leap and crash, making water accelerate and slow, making sediments float, roll, slide, and bounce, or stop on the riverbed for years. Gravity was chaos, yet it had created the patterns and grandeur of the Grand Canyon. In his father's cosmos, the speed of light was absolutely constant, but the Colorado River changed speeds with every curve, thus creating curving cliffs. In his father's cosmos, time served unvarying law, but at the canyon, time wore a jagged, changing face. In his father's cosmos God did not play dice, but the Colorado River was packed with endlessly rolling and floating dice, including boats.

The Colorado River was even powerful enough to get Albert and Hans Albert to respect each other. One of Hans Albert's assignments, in 1940, was to study sediment behavior as the Colorado River was draining into the new Lake Mead. Under the lake's surface, the river continued flowing, far longer than anyone expected, all the way to Hoover Dam. Hans Albert wondered if this was because the river's extra turbidity discouraged it from mixing with the lake waters. He wrote to his father asking about some basic physics of fluids, and his father wrote back with some suggestions, to which Hans Albert replied: "Many thanks for your very instructive letter. It addressed exactly what I didn't know, and helped me a lot." The distance from the simplicity of atoms to the complexities of rivers was not as great as the leap from rivers to the complexities of human love.

Hans Albert could quiet his scientific mind and see the magic in water. Like his father, he loved sailing. One time he kayaked down the Colorado River, in Glen Canyon, for several days. He saw sediments, yes, new sediments in the water and ancient sediments in the cliffs, but he also saw beauty. If he had kayaked the Grand Canyon,

39

he might have seen no math or theory but only a creature daring the act of living, full of senses and thrill and fear and beauty.

THE MILKY RIVER

When you are at the bottom of the canyon your attention is commanded by the cliffs with all their forms and colors, not the sky, which is mostly hidden by the cliffs. But at sunset the cliffs fade away, fade into blackness, become a mere appendage of the sky, which of course they had been all along, solidifications of the clouds and sunlight and stars. At night the canyon becomes the slit of an observatory dome and points your attention skyward. These stars are the same ones that tried to speak to you at home but were drowned out by roofs and city lights and television screens, but here they are naked and plentiful. The cliff edges make it easy to mark star-rise and star-set, by the hundreds, making it clear you are part of a great movement.

From some campsites, in some seasons, you can see our galaxy, the Milky Way. From most campsites people can see only one short segment of the Milky Way, running perpendicular to the cliffs, but in some places the cliffs are aligned with the Milky Way, which aligns the Milky Way with the river, turning the river into a Chartres window onto our galaxy.

Through human eyes the Milky Way seems calm, but the river knows better and sets things true, sets the Milky Way into motion, reveals it to be a river, which indeed it is, a river of stars. The river sets the Milky Way spiraling and waving, dropping into black holes, shooting out supernova spray. The Milky Way feels a mysterious force, the shore, giving it shape. The Milky Way brushes algae into Einstein's hair. The Milky Way is digging the canyon deeper, revealing ever more time and secrets.

In our mythological imagination, humans often saw the Milky Way as a river, including a river of milk flowing from a celestial goddess—the word "galaxy" comes from the Greek word for milk. To the ancient Egyptians the Milky Way was a celestial Nile; to Hindus, a celestial Ganges; to the Chinese, a celestial Ho. In many mythologies the Milky Way was the way dead souls journeyed to heaven. Many peoples noticed that the Milky Way was thinner in winter and thicker and brighter in summer, and since summer was the rainy season that made rivers rise, they concluded that the sky river was feeding water into earthly rivers.

This galaxy-river idea was more correct than anyone imagined, for the Milky Way galaxy had indeed spun itself into rivers. Matter that had sought the center of the galaxy now felt the same gravity pulling it toward the center of Earth, but it had to settle for the ocean. Galactic spirals became the far more convoluted swirls of currents. Gravity waves became river waves. Stars became a mirror excited by stars. The Milky Way was sipped by mammals and turned into milk to nourish babies. The Colorado River was a river flowing from deep nurturance.

They say that if the sky is dark enough, if you are far from city lights and there's no moon, you can see shadows cast by the Milky Way, including your own shadow. One night at the bottom of the Grand Canyon the conditions seemed right. The cliffs did have a subtle glow—was this from the light of the Milky Way? Were owls hunting by Milky Way light? I walked to the beach, hoping the sand's lighter color would better show my shadow. I looked. I was seeking subtleties that only owls or painters would notice. I did seem to see my shadow. I moved sideways to verify it. I was being followed by a Milky Way shadow.

I saw the river glowing, sparkling. I stepped to its edge and saw that it was filled with the Milky Way. I was staring into the mirror of the Hubble Space Telescope, into a microcosm reflecting the macrocosm. I was seeing the cloud of cornmeal tossed by katsinas as a blessing, this by the hand of an unknowing night that required nothing from us but would still welcome ceremonies of gratitude.

The Milky Way was only a foot away, a hand away. I could never touch the Milky Way in the sky. I reached out and touched this Milky Way, sending a ripple through it, as if I were a supernova. I could feel the Milky Way, but when I tried to grasp it, grasp it with a hand evolved to grasp tree limbs, it evaded me. I was just like thousands of years of humans unable to grasp the universe. But the Milky Way did grasp me, wrapping itself around my hand. I was shaking hands with a spiral arm.

Noticing the cursive ripples I was spreading on the water, I took my finger and tried to write my name on the Milky Way, but my name dissolved away. Or perhaps the river was translating my name into its own language. The river was saying that my name was energy, my name was deepest space, my name was galaxy.

The Milky Way invited me to immerse myself completely, to wade right in, to baptize myself in the Milky Way, but I had to say no, you are cold and I am warm, you are wet and I am better dry, I belong to the air and the earth, I am undeniably different from stars.

There were two times when I did immerse myself in the Colorado River at night, at least in my kayak. One time I had paddled upstream from Lees Ferry, into flatwater Glen Canyon, when a monster storm, with gusts that could flip a kayak and foot-high waves rolling upstream, forced me to hide in a cove for hours. I watched waterfalls breaking out. By the time the storm quit, the sunlight, too, had quit, and I had miles to go to get back to the boat ramp. I saw planets and stars appearing in the now-clear sky and in the river. I was paying keen attention to the water, trying to recognize the currents pushing me back and forth, the boils making me wobble. The stars too flowed and fluxed. I planted my paddle into the stars and swirled them further. My paddle became the paintbrush of Van Gogh's sky.

On another occasion, in another canyon of the Colorado River, we were at a final rapid when night came, with another seven miles to go. Once again invisible forces, coming out of nowhere, pushed and spun and tilted me. This time, not being alone, I was better able to relax and enjoy floating through the stars, to admire how the water's darkness blended into the darkness of the sky. I didn't mind feeling very small in the grip of something much more powerful than myself, flowing unstoppably onward. I could feel it then, not just think it but feel it through motion and muscle and balance and joy, feel the energies flowing through the universe, feel the darkness and the light, feel the strange forces steering. The night said that inside myself there was a greater self, a river within, a galaxy within, flowing with a will of its own, upon which I was merely a boat, able to maneuver just a bit, perhaps maneuvering myself into false pride, when the galaxy-river was saying that I was only one drop, one sparkle of itself.

Conversing with the River

Some of our exploration of the Grand Canyon will be conducted by kayak, so it's time to better introduce the kayak. In Western culture kayaks are perceived as a thing of mere sport, with the goal of conquering nature. For the native peoples of the north who invented and perfected them, kayaks are instruments of living, full of humility before nature, full of sacredness.

In one Inuit story, kayaks get the credit for bringing light to a dark world. Usually it is Raven the trickster who steals the sun and makes the world inhabitable. Stealing the sun is necessary because the northern universe lacks a god who lets there be light out of his almighty power and benevolence. In this version, a human builds the best kayak ever made, sets off into unknown seas, and paddles day after day, asking whales where to find light, but the whales have never seen or heard of light. At last he sees a ball of light rising into the sky, then sinking. He follows the light to an island and a house where a man guards the sun by night. When the guardian falls asleep, the kayaker sneaks inside, grabs the sun, races to his kayak, and speeds away. The guardian chases but has no kayak and cannot follow.

The kayak earned its starring role in this story because the Inuit depended on kayaks for survival. They devised kayaks specialized for rivers or bays or open seas, for solos or teams, for speed or stealth in hunting, for the stability to tow dead whales, for carrying families to the next village. They developed the skills and gear for paddling through storms, navigating open seas, and righting a capsized kayak. Kayak paddles propelled life through the generations.

The Inuit and other kayak builders felt a deep bond with their kayaks. It only begins with a person's physical snugness in the boat, worn like a glove, an extension of your own body and balance and desires and motions, merging you with the rolling waters. Kayaks were not just objects but living beings who shared your life and sometimes your death, perishing with you at sea, being laid atop your grave on land. Kayaks were built by hand, carefully and lovingly, out of gifts from nature: bone, seal skins, and wood. The wood was inherently magical because northern lands often had no trees, and kayak builders depended on driftwood found on shore, wood from distant, unimaginable forests. Kayak building was accompanied by elaborate ceremonies, presided over by a shaman. The builders sang secret magical songs that blessed the kayak and empowered it for safe traveling and successful hunting. They made food offerings to the kayak. The kayak owner went naked and sang his childbirth song to his kayak. With long grasses the builders made sweeping motions away from a new kayak, brushing away evil spirits. Launching a new kayak called for ceremonies and feasting that might go on for days.

Throughout the world, seafaring peoples and river peoples often gave boats central roles in their mythologies. Ra might be the Egyptian sun god, but he carried the sun across the sky on his boat. The greatest Greek hero quests, of Odysseus and Jason, were made by boat, and Charon rowed souls to the underworld. Noah was only one of many heroes whose boat saved humanity from a great flood. From the South Pacific to the Amazon, the greatest gift from the gods was the knowledge of boatbuilding and navigation. From Scandinavia to Africa, kings and ordinary hunters were buried in their boats. Many boat cultures placed shrines on board for making appeals to the sea gods, and they held ceremonies to bless new boats, often making offerings of food or wine. Christianity absorbed such rituals and renamed them "christening," a baptism for a new and vulnerable baby.

I admired the Inuit sense of intimacy with their kayaks. More than any river trip in the world, two weeks of kayaking the Grand Canyon bonds you with your boat. The Inuit viewed kayaks as keys opening the door to forces far larger than humans. I could not legitimately bless a desert canyon boat with the song of an Arctic shaman, but I liked the idea of launching my journey with the right spirit, not as conquest but as quest, needing ability but also humility, presenting not just obstacles but oracles. I hoped my kayactivity would place me on the river's wavelength and not be merely about myself.

I wondered what would be a legitimate, natural blessing to a kayak and a journey.

One day I noticed I had been performing a little ritual all along. As I was sitting in my boat, halfway in the river, ready to launch, I was reaching into the water and splashing the deck of my boat. Sometimes I was washing off dirt, but this was hardly necessary, as the waves would soon do this anyway, and I was splashing my boat when there was no dirt. Perhaps I was symbolically saying that this boat belonged to the river and not the land.

I reach toward the river, and the river reaches its shimmering hands toward me and we link hands. Yet we also flinch at our alienness. The river is too cold for my flesh, and it reminds me that a river that can carve a great canyon could erode my flesh far more easily. My hands disturb the river, diverting its smoothness into vortexes, the same monster hands that build dams.

I cup my hands and lift the river. River antibodies rush in and heal the river's skin. I toss the water onto my boat. The water unfolds and flows back into the river, the river not wanting to leave itself. My splashing is very modest compared to what the river will soon be doing to my boat and me, and it is no vaccine either. I am simply asking the river's fearsome power to welcome me, asking the river's clarity to wash away my opacity, to carry me not just into the canyon but deeper into myself and the universe, into not just gravity but possibly gratitude.

Yet my imitating a Christian baptism was problematic. Christian hands raised in baptism are expecting to be embraced by greater hands, something more solid than water, coldness, evasion. But this was the deal on which the river and I had shaken hands. This wasn't about where my offering was arriving but where it was coming from, the gift it makes within. I was raising my hands for the sake of raising them, going on a journey not destined or guaranteed but worth doing for its own sake.

My little ceremony was only the beginning of what I hoped would be a larger ceremony. My real offering was not river water to my kayak but my kayak to the river, myself to the canyon. I was offering my eyes to the blind cliffs. I was offering my ears to the river and wind and birds. I was offering my nose to the trees and mud. I was offering my touch to the rocks and leaves. I was offering my voice to the silence, my joy and fear to the indifference.

I launched, left my element. I floated, a corpuscle in Earth's veins, in a pulse that had become my own, and if I had forgotten my origins and deepest identity, the river would do its best to remind me by accelerating itself into rapids and accelerating my heartbeat into concordance.

Immediately, my kayak started drifting, wandering, as it would continue wandering for the next 225 miles, even when I was trying to direct it. Wandering was rooted in its DNA. The Greek word for wandering is *planktos*, from which comes the word "plankton," the marine plants that are not attached to the sea floor and drift with the currents or winds. My kayak was made of marine plankton that had lived 100 million years ago, part of huge rafts of plankton that died and were buried but still contained the sunlight they had harvested, and one day they were summoned to give that energy to humans.

The summoned plankton was also molded into shapes, like kayaks. Even as a kayak, the plankton retained some of its ancient self. In the carbon of its hydrocarbons my kayak remembered the promise it had bonded 100 million years ago to defend life against the randomness of the ocean, and thus it was defending me now. In the hydro of its hydrocarbons my kayak remembered being the ocean and floating ever onward, and thus it was carrying me onward now.

Drifting, I could feel the plankton renewing its old habit of riding the currents and winds wherever they wanted it to go. Drifting, drifting too far, I directed the plankton to drift back into the current and go straight ahead, and the plankton obeyed with its old compliance to forces stronger than itself, but a lot had changed in the 100 million years it had been hidden underground. Now it was being steered by a creature that could generate its own currents and purposes. The plankton had been content to let random currents add up into the survival of the plankton species, but I took the survival of my decimal of the human species a lot more personally, and thus I had to constantly override my boat's willingness to drift anywhere. I was teaching it the rules of a new game, but it seemed to recognize the same waters on which it had thrived eons ago, on which it now announced, with the spreading arms of its wake, the resurrection and the life.

My plankton kayak was the right vessel for a Grand Canyon journey: as hydrocarbons it had spent 100 million years exploring the depths of earth and time.

When I pulled my kayak ashore and turned it over, I saw hundreds

THE RIVER

of scars from scraping rocks. Some small scratches came from my quietly sliding my kayak in or out of the water, but other scratches were deeper and several feet long, sometimes curving, seismographic measurements of the energy of the current that pushed me over submerged rocks I didn't see or couldn't avoid, also leaving exclamation marks in my brain. These scars were miniature canyons carved by the same power that carved the Grand Canyon. Were my kayak scars like the curving trackways of elementary particles in cloud chambers, revealing the bones of the universe? Or were they a child's coloring book scribblings, a Jackson Pollock painting revealing only chaos? At the least, they recorded my kayak's journeys. If I could read them I would recognize the fingerprints of limestone here and granite there, House Rock Rapid here and Hance Rapid there.

The river was making other marks on me. As I reached out my paddle and dipped it into the water and pulled it back, the water parted into a bulge and wave that grew and spread. I was sculpting the water the way a potter sculpts clay with her fingers and wooden paddle. And with every stroke, the river was sculpting me, gripping my paddle and reaching into my muscles and dripping its strength into them, strength I might not feel day to day when I ended up tired, but which I would notice after two weeks of paddling. With every stroke I did feel the river's strength resisting my paddle, even in calm water, first with surface tension and then a deeper force, revealing that even water had an identity to defend. In stronger currents, the river wrestled me for control, attempted a mutiny on my boat. I was feeling the power that carved away a mile of rock, rock that had intruded upon the river's identity but been pushed aside. In the water's strength I felt the height of the Rocky Mountains, the force of the cascades falling from them, and mountains melting away into sediments. Yet this mountain-devouring god was happy to let butterflies and trees drink from it, and to magic-carpet me through the canyon. Such is the theology of the Colorado River.

After the trip the river's strength would take awhile to drain from my muscles. Weeks from now the Colorado River would carry grocery bags holding California lettuce containing curls of the very water that had curled in Grand Canyon rapids, and when I ate these

47

green curls their water would flow back into the muscles they had made and curl again.

My paddle was wood, its sensuous golden woodgrain holding its own imprint of flowing water. My paddle had begun as ash trees soaking up water that was trying to become creek water and river water but that got summoned into a river flowing into the sky, into a delta of green leaves. My paddle still showed those skyward channels, and showed the annual rings of varying generosities of rainfall. My intercessor with the Grand Canyon held its own stratum of sedimentation, its own fingerprint of time, to which my fingerprints pressed tight—rain and river formlessness now able to grip—clutching at time, confirming my own solidity and yet fluidity. The fossilized water ripples in my paddle flowed back into the river, freeing themselves, filling the river with ripples both ancient and new, ghost water giving its immortal self new shapes and directions. Through this wood had flowed for thousands of days not just water but sunlight and air and earth and, most of all, life—life transforming the water and sunlight and earth and air into the strongest of cells, cells that could support the tonnage of trees, withstanding windstorms and lightning strikes, sheltering eagles and feeding fireflies who at night turned trees into galaxies. My intercessor with the Colorado River remained deeply loyal to life and was ready to defend me from the chaos of water and to shape my passage through it. Tree limbs that had flexed through hurricanes now felt the storm of rapids and flexed but would not break. They supported my entire weight as I anticipated and threw myself into knock-down waves. The confused and shapeless water met wood that had few doubts about its shapes, limbs that became my limbs.

My mostly-ash paddle also held mythological connotations, for humans had long associated ash trees with strength of magical proportions. In Norse mythology the order of the universe was upheld by a giant ash tree, Yggdrasill, which had existed before the worlds of gods and humans and would outlast them. For the Celts the ash tree was a sacred protector and healer. From Greek ash trees came the Meliae, nymphs who nursed the infant Zeus. The Greeks dedicated

THE RIVER

the ash tree to Mars, the god of war, for they made their spears and bows from ash trees. Fittingly, my paddle had been handcrafted by a guy named Homer (if one in North Carolina), so it seemed ready for a mythic voyage.

My paddle held the right element through which to meet the elements, the right beauty with which to approach the beauty of the river and canyon. It was the beauty of a nature that could craft earth and water and air and sunlight into trees. This was the beauty I was offering to the river with every stroke, turning a key again and again until the river trusted that I belonged there and might open its secrets to me.

A kayak offers a wonderful intimacy with the river. You sit inside the river, becoming a river centaur, human from the waist up and a river from there down. Only by thus melding with the river can you fully experience its strength and beauty.

Because whitewater kayaks are designed to spin easily, my kayak reacts to currents I never even noticed and begins to spin, sometimes just a bit, sometimes until I am pointing back upstream, and then, slowly, the river spins me to face downstream again, spins me like a compass needle aligning itself with vast invisible forces.

The river is constantly forming swirls large and small, and in these lines of water wiggling like handwriting the river is telling many stories, if only humans could read the river's language.

One swirl is saying that the sandbar four feet down has just shifted, opening a momentary vortex in the river, which the river quickly fills, except that now the flow pattern above the sandbar is different from before. This swirl is part of the vast flowing of sand by which the hourglass canyon counts off the ages and moves ghost mountains and mesas to the ocean to build future mountains and mesas. As this swirl tugs gently on my boat, it is vast space and time, it is a whole continent, it is the law of entropy and the massive creativity of Earth that is turning me.

Another swirl is saying that for a thousand years a boulder lay in the river and carved up the current. It says that however ultimately strong the river is, for now the river will have to obey rock,

CANYON AND COSMOS

not just boulders but the canyon walls that steer the river's every turn. And so must I. Much of my route is a mapping out of the matrix of tectonic faults and of sedimentary strengths and weaknesses of eons ago.

Within my view there are hundreds of swirls, sparkling as they move, and over the length of the river there are many billions, some of them too feeble to spin a leaf, others strong enough to trap a raft, all of them subtly linked like interconnected gears, like an Escher puzzle. The swirl in front of me is distinctly different because a swirl upstream spun one way and not another; it is one molecule different because a mile upstream a bighorn sheep filled her thirst. If only I could read the river's language, I could read of thousands of events small and large, near and distant. I could see last winter's snow falling in the Rockies, see all the events that carved the canyon, see the evolutionary twists that created harder or softer limestones. And to this dialogue on fate I add my own tongue in the form of a kayak blade, gulping words that will linger inscrutably for miles.

When the river spins my boat to face back upstream, sometimes the boat will re-lock itself into the current and stay pointed upstream for a while. It's quite appropriate for the Colorado River to point you toward where you came from, for this is the vision that no other river can offer so well, a vision of the geological and biological past out of which humans were born, a vision of the enormity of time, a vision of water and sediments and tectonic plates flowing, a vision of flowing life, the strengthening of Vishnu Schist metabolisms into the steadfast heartbeats that became my own, the clarification of a river's roars into a voice that can articulate the canyon's unread and unspoken memories.

My kayak also makes it easier for me to meet the life of the present, for wildlife seems less intimidated by a lone, quiet kayaker than by a raft full of people and noise. I've had more than one friendly staring contest with a bighorn sheep, who would never allow me to come so close on land, and who seems to be curious about river centaurs.

At rapids, my conversation with the river changes tone. The river speaks to me more forcefully, tells me that it is breaking my centaur unity with the river, tells me to obey its hydrological authority. With

50

my paddle I insist on biological imperatives. We are going to have an argument.

I feel the current accelerating and dropping into turbulence. The river yells hard statistics about gravity and volume and the chaotic arrangement of boulders. It yells, if I am hearing correctly, about the people who have drowned in this river, yells that it doesn't care about my life. I reply, less arrogantly, that I understand the river. I understand the push of this current, the strength of the wave that is about to hit me, the true nature of what only appears to be a wave, the patterns amid the chaos. I tell the river about pattern-loving bodies, about the acuity of human vision and judgment, the quick reflexes of human balance, the accurate workings of torso and shoulders and arms and wrists to constantly reshape my body and efforts to the shapes of the waves. I am loyally repeating the river's every word about gravity and volume and boulders, yet I am translating them into biological language. I teach the river of all that has happened on land since the raw waters of oceans and rivers shaped themselves into amphibians and sent wave after biological wave onto shore. As the rapid eases, the currents in my own veins calm down also. The river turns back into a mirror in which I can see myself paddling along.

There is a deep beauty to boating down the Colorado River. It only begins with the aesthetic beauty of cliffs and sunsets. Deeper is the sensual beauty of being immersed in all this. The canyon expands your senses to realms they have never been to before. The deepest beauty is not of the senses but of presence, of being here, here in the grip of a great river, here as a body, getting shocked into a greater sense of reality, here as a strange being on a magical flow.

The Identity of Rivers

What is the identity of a river?

Here in the delta estuary, an osprey diving out of the sky and into the river wasn't always sure about the difference between river and ocean, for the boundary between fresh water and saltwater moved back and forth by many miles according to the tides and seasonal river floods, and the osprey wasn't sure if she was diving into one or the other or some mixture of the two. The freshwater fish and saltwater fish, too, were sometimes confused about when a river becomes the ocean.

Humans define rivers as ending at the seashore, but rivers don't agree. Under the ocean surface, rivers can continue for a long way, even hundreds of miles, in valleys they carved out when the Ice Age ocean was lower and farther away. Because fresh water loaded with sediment is heavier than saltwater, it sticks together and only slowly mixes with saltwater and truly becomes the sea.

Humans define oceans as beginning at the seashore, but then they invent divisions like cove, bay, gulf, channel, strait, inlet, arm, firth, fjord, and seas, and they dissect seas like the Mediterranean into many smaller seas, when the ocean hasn't endorsed any of this and presents the same saltwater everywhere.

For rivers, too, humans may not be seeing real nature but only their own minds, the artificial and arbitrary identities they impose on rivers. Humans seem confused about where rivers begin. From a mountain flows a dozen streams of fairly similar volume, but humans call one a river and the others creeks, brooks, forks, and branches. Some creeks are larger than some rivers. Where creeks merge, humans let some remain creeks while others become rivers. Where rivers merge, both may cease to exist and become a new river, or one absorbs the other. In the Grand Canyon, Thunder River becomes Tapeats Creek, which joins the Colorado River. If rivers are defined by volume, then a creek in flood should be renamed a river and a river in drought should be renamed a creek. Both creeks and rivers sometimes cease being creeks or rivers and are called waterfalls, but only briefly. Rivers can be cases of mistaken identity. Some geographers hold that the Green River is the real source of the Colorado River, not the lesser branch that starts in Colorado. For decades the Colorado River originated at Grand Lake, until a creek that began higher in the mountains got renamed the Colorado River. Rivers never acknowledged any of the names humans projected onto them to give them human meanings. Rivers remained only water flowing, the same water that had flowed through other rivers, over and over. Rivers are not a mere substance, for the water that forms them will soon be a lake or sea, and rivers are not a mere channel, for a dry channel is no river. A river is an

THE RIVER

action, a symbiotic relationship between a channel that creates a river and a river that creates the channel. A river is the presence of endless leaving. A river conjures formless water into shape and power, sends mountains to plow canyons, disguises the ocean and tricks humans into seeing only rivers.

If humans are so illogical about a simple force like a river, then how much can we trust them when they gaze into a river and see a human face wavering there and feel certain they recognize their own identity? Could human faces, too, hold depths and complexities humans do not readily perceive? Could human faces be yet another disguise of the ocean?

Rivers never recognize themselves. They don't see the mountain snows and rainstorms from which they are born, the trickles becoming cascades. They don't see themselves curving back and forth, quickening and slowing, through mountains and canyons and forests and prairies. They don't see the life that lives in them and from them.

Rivers never recognize the work they do. The Colorado River signed its very name as *Grand Canyon* but could never read it.

Rivers never identified themselves as rivers. Almost all the water on Earth is part of the ocean almost all the time. Perhaps one-thousandth of one percent of Earth's water is found in rivers, and none of it for long. Some of the water in the Colorado River has never joined it before, some had been here a million years ago, and some was here only last year. Yet humans look at this mass of random strangers and see only one solid entity: the Colorado River. As the ocean this water was formless, and as the Colorado River this water retained its essential formlessness, so strong it was dissolving the cliffs into formlessness.

When the river flowed into the form of a fish and flowed in new ways, the fish did not perceive itself as the river, and when the osprey grabbed and swallowed the fish precisely because it had solidity, the osprey did not imagine it was eating formlessness.

And then one morning, one more emergence from the unremembering night, the river awoke.

Naked presence. That was what the river saw when it awoke, not

CANYON AND COSMOS

shapes with names, not predictable activities, not meanings disguising things, but the simple existence of a world.

The river opened its eyes and saw a vast sparkling, hundreds of little lights flashing upon some rising and dropping energy the river somehow knew to call "blue."

The river looked up, far up, and saw more blue. Was this the same blue? At that instant an osprey dove out of the sky and into the water, and the river decided: yes, the water is only a continuation of the sky, perhaps the daughter of the sky.

In between the two blues the river saw a massive interruption, many bands of color, loaded with shapes and textures, confusing. The river scanned the cliffs and the sky and the water, hoping for recognition. The cliffs seemed too hard. The sky felt too empty. The river saw the water flowing and sparkling, and felt affinity: *This is what I am.*

The river refocused on something right in front of it, something odd, with five slender appendages. Somehow the river knew that this hand was full of water. The river directed its energy into the hand and, yes, the hand rippled. *This is myself.*

The river raised the hand to touch something full of knobs and cavities. Touching two cavities, the river felt disturbed, weakening, urgent. The river removed its hand and felt air gasping in and out, felt better. What was making all this happen? The river realized: *I have a face.*

Now the river remembered that it was a river and had been a river for a very long time, flowing, sparkling, seeking something, testing out every shape it met, every cliff and boulder and beach, fitting itself into them, taking on their shapes, becoming curves and straight lines and wrinkles and nooks, impersonating the shore not just with shapes but with reflections. And now the river had flowed into another shape and filled it, the strangest yet, these hands, this face, this body, this deeper power of reflection.

Ahead, on shore, stood something blue. It too had a face, but with a long, pointed beak. It was standing on a rock and staring into the water. Was it staring at its own reflection and contemplating the boundaries of selfhood? When the river got close, the heron tensed and flew, flew both over and within the water, with no apparent

54

THE RIVER

disjunction of selves. The river wondered why the heron was fleeing. When the river had been only water, it had flowed past herons endless times. Was the river's new face so frightening? What strange order of things was this in which the river in different forms couldn't recognize itself, saw only alien and frightful identities?

Ahead, the water was going bizarre, roaring, disappearing. The river knew this was a rapid, like so many the river had flowed through for ages without any worry. But now the agitated water was stirring a further agitation. The river had flowed into the disturbing realm of choice and wasn't sure why it should go onward. *But moving is myself. If I cease flowing, I will no longer be myself.*

The river plunged into the rapid, but the waves were no longer itself, only the slap of alienness, threat. Suddenly the river had a right side and a wrong side, an upside-down side urgently disturbing, leaving no further doubt the river had been exiled from its past self of innocent, oblivious flowing. The river yearned for the sky, and suddenly the sky was flowing into the river, and they embraced each other happily.

The river gazed up at the canyon it had carved and gazed down at the tiny body it inhabited, baffled at how it had come to be such a thing, this body more skilled than a river yet more limited, stronger yet weaker, filled with strange needs and questions.

And the canyon, after ages of being only an image wavering in the river, the canyon now wavered inside a mind, tasting needs and questions, seeing itself as haunted, asking: *What am I?*

WATERMARKS

Pressing my hands against the walls of the sinuous slot canyon, I felt the smoothness of the rock, polished by water swirling back and forth, down and around.

Hidden throughout the Grand Canyon, often hidden beyond their bland openings at the river, are side canyons with slots, grottos, pools, waterslides, and waterfalls that defy the canyon's rough, dry, inhuman dimensions, that offer a scale and geo-architecture and fantasy playground humans find delightful. They also require you to focus, like right now to avoid slipping while climbing, on details like rock's texture, striations, angles, colors, and fossils, details that often get lost amid the larger scenery. This limestone was built in a quiet ocean 500 million years ago and now was unraveling with the same hydrodynamics that made the Colorado River and the canyon twist back and forth. Now only a trickle of water was flowing through the slot canyon, but I imagined flash floods thundering through, reshaping its contours.

I was going the wrong way, uphill, against the desire of water. Then one of my feet, pressed against a steep wall, slipped, and I slid down a bit, just like the water. I was reminded that I was indeed mostly water. I felt the weight of water. My water was feeling the same gravity that was calling this trickle toward the river, calling the river toward the sea, and calling sea calcium toward the seabed to become limestone. My water was trying to rejoin its flowing brethren, but I wouldn't let it go. It belonged to me, my shapes and purposes. I willed my water to defy gravity.

Some of my water molecules had flowed through this slot canyon as free water, many times over tens of thousands of years. Some of my water molecules had recirculated through the Southwest, percolating through aquifers for centuries, emerging at springs, being sipped by ravens and bighorn sheep, flying and climbing for a while before seeping into the ground again. Others of my water molecules had gone on global adventures, through oceans, glaciers, numerous great rivers, forests, and every kind of animal. I contained vast history and work, but my molecules had forgotten all of it, leaving me to guess.

I had to invent memories. I imagined my water seeping from canyon springs, drumming from thunderstorms, and melting from snowbanks on the rim above. I imagined my water molecules swirling through this slot canyon, swirling to the Coriolis force, swirling with gravity, swirling with their own momentum, swirling with the curves of the rock. My water was the sculptor of this slot canyon, the creator of a beauty it never saw, until now. Now it has returned, with hands and feet that are eroding more grains of rock, with eyes that can reveal its rock secrets, with a mind that can say: these are our tracks. We helped dig the entire Grand Canyon, and half a billion years ago we helped create this limestone.

THE RIVER

I could feel this Michelangelo water at work inside me now. Flash floods were flowing with tight control and skill, swirling through veins and cells and organs. Erosion was constructing. Water that had died trillions of times was still refusing to surrender.

Yet what right did I have to call this "my" water? True, I could make it grip rock and move upward against its own inclination. I could make it flow down a river in new ways. But it retained its own laws. I had little control over how it was flowing within me, what it was doing in cells and organs. "My" water demanded: *reinforcements, now, thirsty,* and I couldn't refuse. "My" water announced it was done with me and demanded to leave my body, and I couldn't refuse, not for long. "My" water came and went on its own schedule and purposes. It was only visiting, loaning itself to me for a brief time before heading back to formlessness or other forms. It had gone on great adventures before me and would go on many more after me.

And who or what was using the phrase "my water"? I was a vast collage of molecules, primarily water, though water might not be the leader in complexity and guidance. Why should water be calling itself "my water"? Would a river say "my river"? How did water become so alienated from itself? It should be saying "myself." It should be using this body and mind to see itself at last, its long journeys and transformations and work. Water sculpted human faces as much as it sculpted this slot canyon and the Grand Canyon. Yet water seemed to imagine it was merely a human, nothing but one body's own experiences and desires and social roles.

Where in this collage of molecules was the "me"? How were all these molecules merging to make decisions about moving this way or that? Why did they imagine they were only one creature? An orchestra had mistaken its music for itself.

I reached the top of the slot canyon and emerged into a cliff-enclosed stone plaza with pools and chutes of water trickling toward the slot canyon. The rock was full of fossil shapes and erosional shapes. The universe is fond of making shapes, making separations.

I sat beside a pool and watched water flowing out of it, rippling. From minute to minute the water in the ripple was new and the water was disappearing, but the ripple remained, flickering a bit but holding a consistent shape. It was like the waves in the Colorado River, which continued their basic shapes from day to day and year to year though the water that performed them was gone in seconds, gone to perform other waves. It was like a living body, though water paused longer in them, and body waves grew and declined over decades. It was like me. I looked into the pool and saw my face, full of ripples. I perceived my face, my sense of solidity, dissolving. I felt myself flowing. I felt water flowing on its ancient rounds, through river and ocean and storm, and rippling through me. I heard the waves and surf and thunder within me. I felt all kinds of other molecules on their own long journeys through other kinds of waves, pausing within me. I was a rendezvous, a collaboration of molecules loaning their talents to me, and through me meeting themselves, seeing their true identity, even through the "me" carnival mirrors of the human mind. "I" watched myself rippling onward, into the slot canyon, into the rock ripples we had made because we had to, because the universe directed it, because our music was magnificently sinuous even if it had gone forever unheard.

RIVER GODS

To honor Tiyo and his life-winning journey down the Colorado River, a tip of the North Rim was named Tiyo Point. It overlooks the Egyptian section, six buttes and mesas named for the gods of ancient Egypt, including Osiris Temple. Like Tiyo, Osiris made a journey down a great river and thus earned the most important gift for his people.

Osiris was the god of the Nile and of agriculture, a connection compelling to the Egyptians, for the Nile's annual floods renewed the fertility of farmland. Osiris ruled Egypt with great wisdom and success, but this made his brother Set jealous, and Set sealed Osiris in a chest and threw it into the Nile and it floated out to sea and landed far away. Osiris's wife Isis found the dead Osiris and brought him home and performed all the right rituals to give Osiris eternal life, making it possible for all Egyptians to achieve eternal life.

The Egyptians and Hopis are both farmers in the desert, an inherently problematic lifestyle. Yet the Hopis live a long way from the Colorado River, and their survival does not depend on it but on rainfall. Still, the Hopis made Tiyo's river journey central in their religion. The similarity of the Tiyo and Osiris stories reflects the mythic power that rivers stir in the human imagination.

Water being the body and motion of life, the love of water is planted deep in genes, roots, flesh, and minds. When water became human and tried to fathom the world, water poured into our stories about it. Humans filled rivers with gods, some benevolent, some untrustworthy, some malevolent, some demanding offerings. For peoples living along the great continental rivers, relying on them for food and transportation, river gods often became the strongest and most important gods. Rivers wove their way into creation stories.

For the Mandan tribe of North America, the Missouri River is a thin remnant of the primordial seas that originally covered the entire Earth. For the Skagit Tribe, the god Schodelick, after creating the world, went to live in the Skagit River, and Skagits made vision quests there, fasting and diving to obtain some of Schodelick's wisdom. For the Sumu people of Central America the world was created by two brothers who then took a canoe trip down a river to admire their work but got dumped by a rapid. They built a fire to warm up but it burned them up, turning one into the sun and the other into the moon. For the Huron people of Canada the creator gods were twins, one good and the other evil, battling to shape the world: the good twin made rivers flowing both ways at once to make canoe travel easier, but the evil twin made rivers flow only downhill and filled them with rapids.

As human societies grew larger and more complex, tribal and nature gods turned into more universal gods, responsible for human origins, morality, and fate, and river symbolism evolved too.

In the Babylonian creation story, Marduk, the hero of order, slew Tiamat, the dragon of chaos, turning Tiamat's body into the earth. From Tiamat's punctured eyes flowed the Tigris and Euphrates Rivers, still holding too much primordial chaos, requiring people to gather at Marduk's temple to celebrate the triumph of cosmic order and to pray that the rivers not flood and destroy their homes. In the Viking universe all rivers flowed from the base of Yggdrasill, the tree that upheld cosmic order. For the Hindus, Chinese, and Egyptians, their great rivers flowed out of heaven, first flowing across the sky as the Milky Way and then dropping onto Earth.

All over the world, humans gave rivers the powers of spiritual cleansing and bodily healing. Shamans sought out waterfalls to receive visions. River goddesses granted fertility to women. Achilles won immortality by being immersed in the River Styx—except for that heel. When John the Baptist immersed Jesus in the River Jordan and heaven glowed with joy, baptism became one of the central rites of Christianity, maybe essential for salvation. Millions make pilgrimages to the Ganges River, where they immerse themselves in its divinity and send the ashes of their dead back to divinity.

Water could purify not just individuals but the entire human race. All over the world, humans spun myths of gods sending great floods to punish an imperfect and ungrateful world and to purify it for a new beginning. Flood myths seemed most elaborate among peoples who lived beside great rivers, peoples who saw rivers as both the greatest gift and greatest punishment, who felt the tension between gratitude and vulnerability. For the Ute and Mojave tribes it was the Colorado River that caused the deluge. For the O'odham, the Colorado River and its canyons were formed by the tears of a magic baby trying to destroy creation.

The rivers of creation, the sacred rivers, pulsed with primeval power, the power from which flowed the cosmos and human life. They connected the gods and the earth, the heavens and the underworld. They were newborn and eternal, raw and perfect, creative and destructive, giving life and giving death.

A river of creation created the Grand Canyon. The canyon is the trackway of a strong brown god. Like many creator gods, the Colorado River could not separate creation from destruction. To create the canyon, the river had to destroy hundreds of miles of rock. Once begun, this creation would destroy itself, eventually turning the canyon into a wide valley.

For many of the animals living inside the canyon, the cliffs and river and agave are their entire universe. They never suspect there is a different world beyond, a world of plains and mountains, pine forests and bears. Animals quench their thirst with river water yet never wonder where the river came from, never imagine that the river has been flowing for a thousand miles and for millions of years. Likewise, they never wonder where their own lives came from.

Only fairly recently did an animal arrive at the canyon who could look into it and see the powers of creation, the source of their own lives, the strong brown god that even created the gods.

DEEP IDENTITIES

The Grand Canyon can change your sense of reality and identity.

Before people come to the Grand Canyon they have few occasions to see that human society is but a thin film atop deeper realities, that the floors that seem so solid, the stages on which we play our identity games, are only inches thick atop thousands of miles of rock and magma rivers. But on the canyon rim, from some angles, you can see that the hotels that seem so solid and stylish from the inside are but tiny sandcastles soon to be swept away by an incoming tide.

At the canyon, people see the rock layers, the secret strength and complexity and time and beauty, that had upheld their lives all along. Rather than smallness or alienness, people should feel affinity with the canyon. Each tourist body holds the former dust of the Grand Canyon, dust migrated onward, dust now cohering and making a homecoming. Each body holds canyon cliffs that long glowed with the sunset beauty they can finally behold. Sediments that had written a mile-high geology textbook—sandstone cross-bedding now brain cross-bedding—are now able to read their work. Former fish are walking atop their own fossils. The former Colorado River is still feeling gravity but now with feet, with fear of falling.

Yet these canyon ghosts have a hard time being heard through the social identities people bring with them. People are thinking of themselves as teachers or businessmen or mothers, as the roles and status they play within human society. The photos they take are often less about the canyon than about proving that a famous place has made them more important. Some tourists get withdrawal symptoms from their social identities and continue calling the office or home to reassure themselves they have not disappeared.

Humans are intensely social animals, our survival dependent on families and groups and societies, our emotional health dependent on how people treat us. Hermits are assumed to be a bit crazy. The evolution of the human brain owed more to needing to figure out the complexities of human relationships than the complexities of geology.

Yet the canyon doesn't judge people by their social identities. It treats people as only human bodies, only animal bodies, defining soft bodies by hard rock, air-breathing bodies by deep water, water-drinking bodies by dryness, gravity bodies by steep drops, mammalian bodies by scorpions.

As you descend into the canyon by either foot or boat, the canyon becomes larger and you become smaller, the canyon becomes more real and human society becomes more distant. The canyon cuts you off from the constant feedback that maintains social identities. If you are on a river trip lasting two weeks, the canyon can go from being strange to normal to the only world, the overwhelming world. You live more elementally, immersed in river and rock and sand and wind and sun. The night sky is the most real and strong night sky you've ever seen. The canyon redefines the relationship of rock and life: people come here from much greener places, but in the canyon, rock dominates and life has to grab hold awkwardly, green specks here and there. On the trail you may perceive the dust on your boots as an intrusion of chaos onto the proper order of things, but maybe the canyon is covering up your usual identity and telling you about fertility and time. On the river, to the waves hitting you, you react as a furry mammal who likes to remain dry and warm, but maybe the river is trying to wash away your mammalian identity. The canyon is painting your portrait with ancient pigments and watercolors, impressing sunsets onto your face, giving you an enigmatic smile. Starlight is seeing through you like X-rays. Watch closely now. The earth and water and starlight are signing their work. They are writing your own name.

It may not be easy to perceive what the canyon is telling you. There are people who spend two weeks in the canyon and barely notice it, only talking about their favorite TV shows. The canyon has offered me many mirrors I have not seen.

I come to a dripping spring and its small, clear pool, reflecting the cliffs and sky, gently rippling, part of the great cycles of water. This water is offering me a truer mirror than the mirrors in human homes, which show us only how we want to appear to other people, our hair carefully styled, our clothes clean and straight. This water would remind me that my own face is rippling, that when I was a pool like this I had shown wrens their faces. But I don't kneel at the pool as a pilgrim, only as a thirsty animal in a desert.

Another mirror was alive. One time I came across some bobcat tracks leading out of a side canyon and down the beach to the river. I thought only of how different a bobcat and its canyon life were from me. I did not see that she had come to quench the same thirst I felt, that our footprints were different letters of the same sentence.

But if you aren't careful, in a moment of weakness when you aren't projecting your human self or defending your animal self, the canyon can seize you and pull you out of yourself and show you the earth deep within you and the river sparkling in your eyes. I once did a Grand Canyon trip with a longtime Broadway actress, whose entire identity consisted of fake identities, but now she was dying of cancer and she had come here seeking something more real. As the water beat on her raft, as on Shiva's drum, she felt the flexing of creation and destruction.

There's another way in which the canyon can challenge your sense of reality and identity.

At the bottom of the canyon, in the middle of the night, you wake up and you don't know where you are. Your surroundings are alarmingly strange. Such awakenings can happen in hotel rooms where things are only mildly different from home, or even in your own bedroom when dreams have carried you to other places and times. Waking up in the canyon night, the world is even more dreamlike, and perhaps you can glimpse, if only for a moment, that the world truly is strange.

Before you can wake up in the middle of the night and not know where or who you are, you must first fall asleep. This, too, we phrase as a human action, *I go to sleep*, as if we are in charge of it, when in truth sleep is in charge of us, no more stoppable than the turning of the planet and the fading of daylight into night. Consciousness, too, fades away and night takes over our bodies and minds, declaring that we belong to it, to forces far larger and more powerful than ourselves, the same forces that demand we breathe and grow and age and die. The night demands that we lie down in regular supplication to it. In dreams, the night toys with our daytime identities. Daylight allows us to borrow these bodies for another day, parking them in offices and restaurants, using their abilities, but we always have to return them to the twilight zone and to night.

In the canyon you could wake up in the middle of a night billions of years old, wake up in alien bodies with their own purposes. If you are fortunate, your confusion doesn't end after a moment of refocusing, after you think "Grand Canyon sand," but continues to tell you that the world and you are surrealistic. The strangest thing about humans may be that we don't find the world and ourselves to be strange, that we don't spend all day in a state of midnight startlement.

WHO?

In the canyon night the universe called out to me and asked me what I was.

Actually it was an owl calling out, but in its hiddenness, in the echo chamber of the cliffs, this call could seem a greater voice. *Who are you?*

I turned in my sleeping bag and tried to focus on the call, but I could not tell where it was coming from. *Who are you?* the night insisted.

I was tempted to answer, but what could I say? Among humans it would be enough to give my name or my role in human society, yet neither the owl nor the night would understand such identities. I was in their realm and needed to define myself in nature's society. Would it be enough if I identified myself as a human?

Who? said the owl, as if already rejecting this answer.

I was in the zone between being awake and asleep, where logic sags into dreams, and thus it seemed important to justify myself to this call in the night. I could have told the owl that humans are a primate that came from Africa, but a Grand Canyon owl would know nothing of primates or Africa, and would answer only "*Who?*",

THE RIVER

demanding something more. I am life, I could say, the same as you, only in a different package, with arms instead of wings, but the owl would answer only "Who?" and I would be forced to retreat through the whole evolutionary story, backward like a child listing his address as Earth, Solar System, Universe, back to the creation of the universe, and still the night would insist: "Who? Who?" just as humans have always stood before the mystery of creation and asked "Who?" Was the owl only wooing me back to a final agreement that my ultimate identity remained a question?

Of course, it was only a coincidence that I happened to speak a human language in which an owl's call connotes an inquiry into human identity. I was turning the owl into an oracle, just as humans have always imposed on nature meanings that nature never intended, seeing only the ghosts of our own haunted minds. But if you listen carefully—and the Grand Canyon's deep silence and magic allows for deep listening—you may just hear that the voice of an owl is the true voice of the night, revealing the final truth about human identity. I am a whoman.

"Who?" said the owl.

I remained silent, and I listened to the night.

The night said that I belonged to it.

Like Stonehenge, the canyon's cliffs and boulders watched the sky, watched the patterns of sunsets and moonrises, the wheelings of the stars and planets, the longer patterns of eclipses and comets, patterns so much larger than human lives that humans often didn't recognize them but which by the canyon's scale were as frequent as sunrises to humans. To the canyon, Halley's Comet was just another migratory bird. The canyon Stonehenge watched the sky with depths that answered the sky itself.

Drifting off to sleep, I watched the stars rise over the rim and sweep their light through the canyon depths.

Who?

Somewhere in the canyon or on the rim were stones that had been arranged by human hands to watch the sun, to welcome the solstice. The hands that had built this altar to the sun also toiled to turn the earth into corn ears they peeled open to reveal hundreds of round

CANYON AND COSMOS

idols of the sun, solar pills they swallowed to infuse the sun's power and sunset beauty into themselves so that their faces would go on glowing through the winter nights and power corn-stalk-holding ceremonies to the sky. The sky had betrayed their loyalty by giving too much sun and too little rain, forcing them to abandon the canyon, and those canyon-soil hands had dissolved into soil elsewhere, but their stones had continued watching and worshipping the sky, projecting the solstice onto ruins or spider webs, in sad, exiled loyalty to the glimpse the stones had been given of a state of existence beyond that of rock.

All night long the stars rose and played their light down canyon walls that had not been arranged by hands or minds and that did not recognize human purposes, did not recognize the lions and hunters some ancient Mesopotamian storytellers had projected onto the sky, did not deny the real stars.

Thousands of rock faces stared at the stars with total concentration. Like surveyor's equipment, cliffs and buttes and boulders measured the angles and alignments of the stars. A star rose directly over the tip of a butte or peeked through a rock arch or shined on a slot-canyon waterfall, but maybe only once a year. Every night the rocks charted Earth's motion through space. And the stars were charting the canyon, charting tectonic faults lifting cliffs upward and the river carving cliffs away. Yet the canyon endured well enough to chart the faults and the rivers of the sky, to watch stars drifting through the galaxy, passing one another, warping the constellations, stars dying and stars being born, and galaxies moving outward.

The canyon watched the sky with fossilized trilobite eyes, with pillars aimed like telescopes, with potholes of water, with owl eyes. It watched with the large white flowers of datura, which opened at night to declare the night sacred, and it watched with flowers and eyes that closed at night to declare the night profane. With the horns of bighorn sheep it calibrated the horns of the moon and welcomed the horned galaxies. It worshipped the starlight that fell into the Sipapu and into the green ceremonial chambers of leaves.

And far into the night, the owl called out "*Who*," and the starlit cliffs echoed "*Who*," and the human, whether oblivious in sleep or rational by day, had no answer.

The canyon and the night sky understood each other fairly well, for they could see themselves in each other. The canyon manifested the same strata of time visible in the sky. Galaxies 275 million light-years away shined on Kaibab Limestone seashells that had been alive when this light was born. Fossil light from stars that had ceased to exist shined on the fossils of species that had ceased to exist. Galaxies 1.8 billion light-years away in the Vishnu night wrapped themselves around the sensual black curves of the Vishnu Schist, and they whispered about the sensual forces of creation. Or was that the river whispering about creation? Or was that the Owl Nebula calling out to its daughter?

CANYON SKY

The sky was flowing into me. The sky's vastness and power and motions and silences, the molecules ancient and new, the breaths of pine trees and cacti and mountain lions were flowing into me and moving me.

Hiking out of the Grand Canyon teaches you about air. Somehow we normally barely notice something that keeps us alive, alive not from day to day like food and water but from four seconds to the next. Hiking makes air real and powerful. Your body turns air into a god and calls out to it. The closer you get to the seven-thousand-foot rim, the thinner the air becomes and the harder you have to breathe. Park rangers like to claim they never stop to catch their breath, only to admire the view. When I stop to admire the view, it includes finally seeing the invisible sky, feeling it pushing through my nostrils, feeling its weather in my lungs and heart and mind, seeing, when I begin to move again, the global winds now moving me along.

Swoosh! The air raced past me, in the form of a raven. The raven called out: *Sky! Sky! Sky!* The raven was pointing out a puzzle about the canyon. When I looked up I saw I was half a mile underground. When I looked down I saw half a mile of sky beneath me, with the raven dropping into it. The sky is supposed to be above us.

Approaching the Grand Canyon, you can travel along flat terrain for fifty miles and then abruptly perceive through the trees ahead a problem with reality, the end of land and the substitution of sky, a whole mile of sky beneath you, a dozen miles of sky in front of you. The birds alone define the canyon as sky, and sometimes the canyon holds clouds condensing right in front of you, rain falling beneath you, rainbows arching between buttes like strata transfigured, and occasionally in wintertime a temperature inversion traps and generates clouds until the whole canyon is a bumpy lake of clouds. This Alice-in-Wonderland sky contributes to people feeling that the canyon is unearthly. Emptiness becomes a true, magical presence.

I was looking over the raven's shoulders as its wings stirred the air, its black quills signing the air into cursives that flowed toward me, so when I breathed this air, it was no longer simply air but the energy of a raven.

Subtly, every raven wingbeat changes the pulse of the world, grabs a batch of molecules out of the realm of physics and rearranges their locations and motions for the entire future of the universe, sending them onto geological multiplication tables and biological roulette wheels, initiating or erasing parallel universes full of ghostly humans.

The raven sent my way not just wing energy but molecules it had breathed in and changed and breathed out, air that flowed into the raven's heart and blood and wings and made the raven fly, air giving itself a new form of motion, as if the raven was merely an enzyme by which the random air became a helical breeze. I soon inhaled a bit of this air, but since it was now mainly carbon dioxide and nitrogen my lungs had little use for it. But my rejected carbon dioxide was given a warm welcome by plants, whose own green wings soon flapped with raven energy. The plants poured out oxygen, a return gift to ravens and humans and other animals, oxygen that moved our wings and legs and thoughts, the ravens now flapping green leaves, the humans now thinking with green flames.

I could feel the energy of plants, immobile plants, pushing me up the trail. Many of these plants were right here, all around me, but did I see them or offer them any thanks? My breath was the breath of ponderosa trees on the rim, algae in the river, cacti I regarded as enemies, wildflowers I valued only for their looks, and lichen so old it had nurtured Puebloan ceremonies here. My breath also came from plants across the continent and the world, plants I couldn't name or picture, plants with millions of secrets, plants dedicated to their own mission, plants oblivious of me. I felt millions of lives pulsing within me. I felt the whole Earth lifting me upward.

My breath was also full of time. Some of the oxygen atoms I was inhaling had lingered in the canyon for a long time, as part of bones, the soil, or the iron in cliffs. Some of it, once part of clouds, may have inspired a Puebloan to paint a cloud pictograph. Some of it had come into and left the Grand Canyon many times. Some of it had thoroughly explored the world and taken part in human history, igniting the first human campfires, proving to Joseph Priestley the existence of oxygen, inspiring Vermeer to capture air's beauty. My breath was also full of animal history: it was roared by dinosaurs and whispered by butterflies.

My breath atoms were also the heroes of geological sagas, of an atmosphere evolving from volcano sulfur to carbon dioxide to oxygen created by life. Out of tiny cells had flowed an entire sky, which inflated millions of types of bodies, the sky turned outside in.

Grand Canyon cliffs contain ancient skies: light rays that had bonded molecules together; rain that had made bodies plump and flowing; winds that had pushed lizards toward becoming dinosaurs; rainbows that had excited eyes, now fossils. The cliffs contained billions of sunrises and sunsets, long separated from the sky but now returned.

When I emerged from the trail I saw people gathered along the rim, waiting for sunset. The Grand Canyon is one of the foremost places on Earth where humans gather every evening and every morning out of an expectation they will see the sky do something special. During the day people are inclined to attribute the canyon's colors to the rocks. Yet people paying longer attention notice the colors changing all day long, for these colors belong to the sky as much as to the rocks, and as sunset nears, the colors change more noticeably and dramatically.

The red cliffs grow redder. Other colors come out of hiding. Clouds too are changing colors. Light beams play along the cliffs, searching for something. The sky wraps everything with sunset. Sunset breezes rise up the cliffs and run their fingers through people's hair and flow into nostrils, and the very molecules that were turning the sky red now become red blood and paint a portrait of the red sky, a self-portrait.

The sky itself is standing here. The human body is made more of oxygen than anything else, nearly a hundred pounds of oxygen, most of it arriving in the form of H_2O rather than as breathed air, but for much of its career this oxygen had flowed not as water but as sky.

I watch the sky within myself. I inhale the sky and feel it infusing me, energizing me, becoming me. In my water I feel the sky solidified, the winds flowing very gently, the rainbows serving as templates for confused colors. My abstract scientific knowledge turns into memory, into me-mory. The sky is filling me with depths that had been there all along. I cannot converse with the sky like a raven, seeing all its details and currents, but only through humans can the sky recognize itself and autograph its deep story.

My normal human identity, which is full of antibodies against being invaded by alien entities, objected to this proceeding, objected that taking breaths out of the sky doesn't authorize me to impersonate the whole sky. To this ego I replied that the sky is generously lending you its strength for a while and you can't treat it as nothing but *your* power. We're talking science here, and it is time you took science seriously and personally and recognized that the identities it gives you are more real than the social masks we wear and take so seriously that we need to occasionally hide them behind katsina masks that remind us that the sun and rain and wind can indeed walk on two feet into our village plaza.

As the sun and horizon touched, I was reminded that my oxygen was born inside stars and swarmed in them in what seemed like total chaos but which held its own weather patterns. My oxygen had built up inside stars until it helped trigger supernovae, exhaling itself into space. My oxygen had joined a nebula until some galactic raven's wing sent the nebula condensing into a new star and into Earth. Every breath I took was the continuation of a supernova, and of the vastly improbable exhalation of the universe, the Big Bang.

THE RIVER

To repay this gift I offered myself to the sky. For billions of years the sky performed sunset with no life to see it, performed it for mountains and seas, performed it among the numerous rituals by which the cosmos affirms its belief in order, declares that order is beautiful. As if sunset was too beautiful to never be seen, the cosmos generated creatures to see it. And here we stood, along the canyon rim, bodies of the sky, bodies made of suns, honoring the sky that feeds us and the sun and past suns that made us.

The sunset flowed into us and glowed with deeper colors, glowed differently in every mind, being filtered by different memories, ideas, and emotions. Some people saw the sunset aesthetically, as colors and shapes, light and shadows. Some people saw it technologically, letting their cameras do all their experiencing for them. Some skulls were Plato's cave in which the ancient sky imagined itself to be merely human. Other skulls were Lascaux caves into which the sun crawled to perform a ceremony of recognizing larger connections.

I offered to the sky and sun a cave wall on which they discovered they now had eyes for beholding themselves, sky-fingerprint ears for hearing the wind, a tongue that allowed an empty bell to finally ring out, and a mind in which oxygen could recognize its birth and its long journeys and creativity. The sunset flowed through me, illuminating the sky within, finding gratitude for a sun that has nourished Earth for billions of years, gratitude for the creativity glowing in the cliffs, gratitude for stars that created oxygen and carbon and everything else we are. The sunset noticed in me what it had been lacking in its ancient ceremony, a celebrant to teach it how to praise. It tried to speak through me: *With this ceremony, with this procession of light and color, against the darkness before the Big Bang, against the eternal darkness to come, we offer this celebration of light.*

Onward flowed the sunset, onward the parade of colors, over the horizon, onward to become Pacific island sunsets and Himalayan sunsets. The colors faded into deep blue, and the humans fled for safe cocoons of light. The blue faded into black, the sky dropped its katsina mask and revealed the true, greater god it had been all along. The stars de-claired: just as you've barely noticed the air that gives you life, maybe you've barely noticed you are alive.

A Home of Rocks

Amid the thousands of rocks around me, a dozen rocks caught my attention, for they formed a straight line. Geological forces seldom arrange straight lines. I was seeing the ruins of a house people had built a thousand years ago. The rocks were the right size for a person to carry, and the right shape to stack into a wall. Little of the wall remained, only three rocks at the highest; most of it had collapsed long ago. The wall sat atop a small hill that had offered breezes in summer, sun in winter, and better views of crops growing, children playing, and sheer beauty. Nearby was a modest creekbed, now quite dry.

The wall was made of limestone, a contrast with the ground around it, a greenish black shale. Shale wasn't the best building material, so the Puebloans had gathered rocks that had fallen off the nearby Redwall Limestone cliffs and smashed into chunks, gathered the world's greatest proof of the power of erosion and argued against it by stacking rocks back up into little cliffs.

The Redwall Limestone consisted of calcium carbonate, the stuff of bones and cement, the stuff of columns that uphold Greek and Egyptian temples long after the gods who lived there died. This calcium carbonate had served as the bodies of crinoids, corals, snails, clams, nautiloids, sponges, and fish. In the rock wall I looked for fossils and saw some vague shapes that implied the work of life and not just particles. Simple and quiet creatures who had sought for nothing but to eat today and live another day had made imprints that had lasted 340 million years, had created rocks and a great canyon and the houses of creatures they had never dreamed of in their philosophies.

The humans who built their houses out of fossils probably wondered about the shapes and color patches in their walls. Did they connect fossil spirals with the snails they saw in their gardens? The Native American mythic imagination was good at transformations, at animals or spirits turning into rocks or mountains, but it was not nearly as bold as the forces and transformations of the geological cosmos.

The Puebloans had lived inside ancient lives. Clam shells kept coyotes and the wind away. Sponges soaked up their words. And ancient lives had lived inside them. Calcium that had defended the lives of clams and snails became their skulls. Shark teeth became their teeth and devoured more fish, whose heartbeats had long ago diverged toward human heartbeats. A human body was indeed a vast reincarnation of past lives. The forces that built these rocks became hands building pottery.

Near the ruin I found a potsherd whose corrugated bands had been created by the potter pressing her fingers into the clay, leaving the impressions of her fingers and fingernails. Her fingers had dissolved long ago, but these fingerprints endured. I touched my finger into her finger imprint, and it fit. I was touching someone across a thousand years. I could feel her satisfaction at touching this clay and giving it shape, at quenching her thirst or hunger from her pottery. I shared her distress when this pot fell and broke. The canyon and rims hold millions of potsherds, the jigsaw pieces of unfixable faces.

I am often amazed by how widespread in the canyon you can find Puebloan ruins, rock art, potsherds, and projectile points, sometimes in unlikely places. A thousand years ago the canyon was full of small villages, perhaps a hundred sites along the river corridor alone. People on the rims could see tiny rooftops far below. The night was streaked with campfires. The canyon was full of trails connecting villages, fields, granaries, creeks, and rock art panels, connecting human lives.

The Puebloans had not just turned rocks into homes but turned the entire canyon into a home. They had not only filled the canyon but been filled by it, not only lived in it but lived it. They were born here, grew up playing canyon games, grew Redwall muscles, fell in love to a canyon moon, gave birth here, grew old watching agaves growing, and died here. Their bodies were molded from canyon soil and plants and animals, were buried in it with ceremonies seashells never gave themselves, and were reincarnated as cacti and coyotes. Their blood flowed out of springs and creeks and the river and flowed into them and swirled calligraphically before returning to the river. Their toil turned the earth into food, turned hardness into clothes, turned geology into art. They turned a blind landscape into conscious maps of where to find passage and water and shade and clay, maps of where and when plants grow and animals move, maps of when constellations rise over a cliff. They turned the canyon into stories that made connections between people and rocks, plants, animals, and stars, stories that made more sense of life and death, that made them feel more at home in the canyon and the universe.

Ruins and potsherds change my feeling about the canyon, making it friendlier, more human. My admiration of sunsets and hummingbirds was an ancient torch that had been passed to me. I could overhear voices in a language that was unfamiliar yet still my own, speaking to me.

It is easy to perceive the canyon as inhuman and unfriendly. While hiking, you have to define rocks in the trail as dangers that could trip you and hurt you. When thirsty, rocks offer no refreshment. You can understand why numerous poets made rocks the symbols of human or philosophical wastelands. But I had come here to ask if rocks could be a home. Seeing some of those rocks turned into houses offered me a greater sense of belonging to the universe of rocks. I could not belong here in the same way the Puebloans had, yet I could see things they never had, including stars piled carefully into a home galaxy.

The cliffs around me had always been at home here, in ways living creatures could never understand. They had been rooted here for eons, never worried about change. They did not beg to be defended against gravity or weather, to be resurrected or reincarnated. It was only when this rock became alive that it felt itself to be a stranger and longed for the cliffs and cosmos to reassure it of its value. It was only through living hands that this rock walled itself off from itself, walled itself off with stone, identities, yearnings, and gods.

Watching sunset, the colors faded away and the stars emerged, framed by a long, jagged window of cliffs. I was deep inside the house of earth, being bedrocked to sleep. Nearby slept the dust of the Puebloans, returned to their true identities, stripped of illusions, embraced by their fellow rocks. As my identity began relaxing its Vibram grip on this body and against rock, the rock drifted closer and claimed me as its own. Rocks disclosed that they were really time and wind and river and stars solidified. Fossils sprang back to life in me. I was feeling more at home in a universe of rock. This was the house that had built us, and through us, by our stacking chaotic rocks into houses, by our shaping chaotic soil into plants and pottery, by our turning the chaotic sky into stories, this house got an inkling of what it felt like, how it was lived, to be at home.

THE BOOK OF MEMORY

I was stopped by the puzzle of memory. I stopped in the middle of a large patch of prickly pear cacti, hugely colorful with flowers. I could not remember there being any flowers when I'd hiked through here a few hours before. Had I already forgotten them? Or had they bloomed after I'd passed by?

Perhaps I was already tuned into the subject of memory because of the T-shirt I was wearing. A few months previously I'd attended a ranger campfire talk at Phantom Ranch, the guest lodge and campground at the canyon bottom, and the ranger, to prompt her small audience into friendly introductions, did a T-shirt reading, asking people to explain their T-shirt themes and thus who they were. She clearly approved of T-shirts from other national parks, but her aloof response to the person wearing a Las Vegas casino logo made it clear that this was an uncool fashion statement for a Grand Canyon hike. I had never paid much attention to hiking fashions, and fortunately I hadn't chosen anything offensive that night. But having been put on notice that the canyon might be judging me, I became more careful. Today I was wearing a T-shirt from the Marcel Proust Marathon Reading at the New French Café in Minneapolis. My shirt featured two famous, elaborate sentences from which *Remembrance of Things Past* takes off. The second sentence:

> But when from a long-distant past nothing subsists, after the people are dead, after the things are broken and scattered, taste and smell alone, more fragile but more enduring, more unsubstantial, more persistent, more faithful, remain poised a long time, like souls, remembering, waiting, hoping, amid the ruins of all the rest; and bear unflinchingly, in the tiny and almost impalpable drop of their essence, the vast structure of recollection.

I took to wearing this shirt on river trips. In the camp kitchen, with the scents of spaghetti sauce and garlic bread welling forth, people always noticed and read my shirt. I explained that this quote was appropriate because canyons are all about remembering things past.

The Grand Canyon is Earth's most vast structure of recollection, the deepest memory bank. It is loaded with nearly two billion years of memories in geological form, and older memories in other forms. It presents its memories in clear lines and colors for all to see. It remembers events as large as continents forming and as small as sand grains rolling. It remembers days of calmness and days of earthquakes and floods. It remembers mountain ranges and rivers that long ago ceased to exist. It remembers a nebula becoming Earth and crystalizing into rocks; in its atomic structures, it remembers the heat of vanished stars and supernovae and the Big Bang.

In its foundation rocks, the canyon remembers the one-celled foundations of life, and, rising higher, it remembers the rise of life. It remembers ancient oceans and their trilobites, corals, and fish. It remembers how strange the land felt to plants and amphibians. It remembers the glory of the sun to reptiles and insects. It remembers life's billions of days and nights, the changing seasons, the changing climates, the ever-changing forms of life. It remembers not just species but individuals, how one lizard went for a walk on one particular morning. It remembers not just with fossils but with the lives in the canyon today, whose genes and bodies and thinking remember not just their species history but the whole history of life. It remembers the changing modes of memory that generated today's novels and T-shirts about remembering things past.

I gazed into my biological pages, my cellular strata. I gazed through my Kaibab and Redwall and Vishnu. I saw memories everywhere, ghosts everywhere, time everywhere, everywhere time moving in new and strange ways. Time was no longer just physics that could be expressed only mathematically. I felt time alive in me, time warm and plump and breathing, time moving and feeling and thinking, time recognizing itself. Time no longer flowed steadily but felt itself to be moving too slow or too fast, becoming tyranny and boredom and impatience. Time looked backward with fondness or regret. Time looked forward with dreams, worry, and plans. In humans, time became the structure for organizing days, for planting and

THE RIVER

harvesting crops, for hunting and gathering, for school and work, for giving birth and raising children. Time became campfire stories about time, about animals and humans learning the lessons of time. Time became novels and songs and plays and movies in which time was the invisible yet dominant character, making all the other characters dance and age. Time became houses and cemeteries haunted by time. Time saw its patterns in sunrises and sunsets, in the cycles of the moon and sun and stars, in the cycles of the seasons, in birth and growth and decay. Time tried to measure and even control itself by building sun clocks and sand clocks and water clocks, by moving metal gears and pendulums and atoms. Yet all along, time recognized, even as humans tried to deny it, that time lived within human bodies only for a while, never even pausing, then moved on without us, into new bodies and adventures. Time held ceremonies to honor time and trick time and stop time. Time invented time machines to break its own rules. Time didn't stop. Time was enthralled to discover what it had been doing all this time, how old and powerful it was. Time saw itself in rock strata and sky strata, in canyon fossils and DNA and cactus flowers. Time saw itself as creative, if only by giving natural forces time to work. Time also recognized it was destructive, killing every life and most species, killing cliffs and planets. Time turned into gods through whose words or wombs time poured into the world.

I gazed at the cliffs heavy and tall and strong with memory. I read the book of memory. It reminded me that I too was full of memories, only a few of which were my own lived memories. I was full of time, every interpretation that time had become in humans, biological time and personal time, scientific time and mythological time, Mayan time and Egyptian time and Christian time and Hindu time, cyclical time and creative time, karma time and liberating time, aimless time and purposeful time, smug creation-story time and endless time.

I take a step. I am time moving. The same time that counts itself by the expanding universe, by clockwork galaxies and solar systems, by biological clocks and springtimes and migrations, I am time moving, the same time that raised these cliffs, I am time moving. Moving into the future. I think about my next step and I am time planning itself, after eons in which time merely happened. I look, and I am time seeing. I am time seeing the vast structure of recollection and the fleeting glory of desert flowers.

STROMATOLITES

It was a mile and thirty minutes to walk, yet it was nearly a billion years away.

From the river we walked up Carbon Creek, walked around rockslides and through a sandstone narrows, and emerged into a fault line that had bent the horizontal sandstone into vertical and left an open valley. We also crossed a biological fault line, the boundary between the Precambrian and the Cambrian, when life went from long simplicity to great complexity and variety.

Amid the shale that indicated this had been a shallow seabed near shore, there were odd-looking boulders, several feet tall, rounded and convoluted in a way unlike other boulders. They tend to remind people of brains. For some 800 million years these brains have been thinking about the evolution of life.

These rocks were the first in the world to be recognized as Precambrian fossils. When Charles Darwin died in 1882 he was worrying that the fossil record didn't support his theory that evolution happened gradually, for Precambrian rocks seemed to hold no fossils and then, abruptly, Cambrian rocks held a great variety of life-forms. Later that year geologist Charles Walcott spent months roaming this section of the Grand Canyon, found abundant fossils in Cambrian rocks, and found these corrugated boulders in a rock layer he eventually realized was Precambrian.

These boulders are stromatolites, once cities of bacteria, now Pompeiis with streets and foundations and mosaics still intact. Stromatolites, communities of many species of bacteria performing different tasks, arose some three billion years ago and became the dominant life-form for two and a half billion years. They filled shorelines for hundreds of miles.

Stromatolites helped to create me. When stromatolites emerged, Earth's atmosphere held no free oxygen, only gases like sulfur and carbon dioxide. It was the cyanobacteria (blue-greens) in stromatolites that developed photosynthesis and devoted themselves to

THE RIVER

converting the air into oxygen, one tiny bubble at a time. This took a very long time: by the era of these particular stromatolites, oxygen was still only a small portion of what it is today. Eventually oxygen became the abundant fuel for animal life, but ungrateful animals began feeding on stromatolites, forcing them to retreat into a few coastal areas so salty and warm that other creatures can't stand it. They are still hiding there today.

People who visit living stromatolites seldom find them attractive. They are dark brown blobs with pocked skin, slimy to the touch. For humans, "slime" is creepy, an insult. Bacterial slime looks and smells and tastes bad, makes people sick and die. Bacteria is an enemy to destroy with soap and medicine.

I reached out and touched a fossil stromatolite. Its slime was long gone, and now it was hard and abrasive, and still mostly brown, not aesthetically appealing. It remained a monument to bacteria, alien life from another planet. How could I relate to a stromatolite?

As I was touching and looking at the stromatolite, my lungs and heart, without my directing or noticing them, were winging the sky into me and throughout my body. In my brain oxygen was drawing pictures of the stromatolite, analyzing its texture and color and value, classifying it with dinosaur bones and other very old and odd things. The oxygen projecting these images and thoughts had been created by these very stromatolites 800 million years before, had roamed over Earth ever since, roamed through numerous creatures and thoughts and joys and alarms, recently roamed through canyon lizards who saw this stromatolite mainly as a good place to sit and soak up the sun. Through its oxygen ghost, this stromatolite lived on. From across the eons, this stromatolite breathed into me, mouth-to-mouth resuscitation, resuscitating itself within me, passing its life into me. Through my eyes the stromatolite saw itself sitting on an ancient shoreline, sparkling with sunshine and photosynthesis.

With my help, through my eyes, ancient and oblivious oxygen studied the egg from which it had been born.

I focused and felt the oxygen flowing through me. The Precambrian wind that had stirred no leaves or wings was now following the convolutions of my brain. The wind was nudging me, trying to

twirl thoughts. The wind showed me some of the thousands of air
and wind gods it had been cast as: Aether with a Greek face, How-
chu with a Chinese face, Quetzalcoatl with an Aztec face, Shu with
an Egyptian face. Air had become the breath of gods blowing life
into clay, and the word of gods creating the universe out of nothing.
Humans had gotten it right: air had breathed life into us. Humans
had given thanks by turning air into ceremonial drumbeats and
chants and songs

Yet it seems God was a stromatolite. How could I say thank you
to a stromatolite? This shouldn't be hard for any breathing human,
and it should come even easier for a kayaker. We are playing a game
whose entire goal is to remain in the realm of oxygen, a goal we seek
with every stroke, every wobble assessment, every balance adjust-
ment, a goal never considered by hikers, climbers, skiers, or ball
players. When the river flips us upside down we feel deep primordial alarms going off, and we are forced to overrule our breathing,
to not breathe water, and we feel our air running out, our despera-
tion for air, and when we roll back up we gasp our love of air. Still,
we tend to take it for granted that air is there and we don't thank
the gods who created it.

For once, I focused on my breath. In and out, in and out, fresh
and stale, good and bad, in and out every four seconds, every four
seconds. I reached out to the stromatolite and felt the vents from
which had come my oxygen, my life. I was doing what humans had
always longed to do, touching not just an altar to a longed-for god
but touching god itself, solid and real. Real yet alien, a strange face
looking back from the mirror. Inspired, I breathed in.

TRILOBITES

I was walking on the seabed, perhaps twenty feet under the surface of
the sea, leaving my footprints in the soft seabed mud—or what had
become of it after half a billion years, the Bright Angel Shale. Being
compacted of mud, the Bright Angel Shale dissolved more readily
than other rocks, back into dirt or mud that took better footprints. I
paused and studied some tracks crossing the trail, lizard tracks with
a dragging tail, and mouse tracks, tracks soon to blow away.

THE RIVER

Then I came across some tracks as old as the Bright Angel Shale. On a rock outcrop beside the trail I spotted some ripple marks, faint and narrow, two sets of ripples side by side, bending with the same curve, traveling maybe six inches. I had seen slabs crowded with such tracks: the footprints of trilobites. The imprints of a few moments of living.

Trilobites, arthropods with three lobes and about two dozen little legs per side, were one of the largest streamers of the Cambrian explosion that filled Earth with complex life half a billion years ago. Trilobites lived all over Earth, generated some seventeen thousand species, and lasted nearly 300 million years. Their species included many sizes and shapes and lifestyles. They lived in ocean depths and along the shore; some crawled the seabed and some swam; some fed on seabed nutrients and others hunted, even hunted other trilobites, leaving two sets of fossil tracks that intersect and one that disappears. This seabed had the right conditions for trilobites to leave fossil tracks, its mud taking detailed imprints that soon filled with sandy mud flooding from nearby rivers. This particular track recalled not just one trilobite's walk but some afternoon thunderstorm.

Where was this trilobite going? Was it out for a morning warm-up, searching for food or a mate, or heading for its favorite snuggle spot for the night? Was it exploring new places, just like me? Did trilobites spend their whole lives in a small, known area, or did they roam continually? This trilobite's life, in its trilobite way, must have been different from other trilobite stories. We were two creatures out for a walk in the same place and crossing paths, just like me and the lizard, me and the mouse.

Where was this trilobite going? Far into the future. These tracks were made relatively early in the 300-million-year run of trilobites, on their way to the golden age of trilobites. The top layer of the Grand Canyon, the Kaibab Limestone, recorded their final forms and diminishing numbers. I gazed up at the canyon cliffs and saw the whole thing as the trackways of trilobites, the story of trilobites, the endurance of trilobites. Now I was the small one under the foot-step of time.

Where was this trilobite going? It was going to turn into myself.

CANYON AND COSMOS

Not directly on the biological family tree, yet if we define life not by its different shapes but by its shared pulse, define it as an ecological knot tying us all together, define it as a long journey through time and forms, then trilobites are a different form of ourselves and their tracks are our own journey.

I looked at where the trilobite tracks reached the edge of the slab and I imagined them going onward. But the archivist rock had crumbled away and these tracks had turned into soil on which lizards and mice and humans planted new tracks, soil into which plants had planted their seeds and turned ancient trilobite shapes into new shapes, nibbled by lizards and mice. Trilobite skills had turned into new skills. Ocean legs turned into heavier legs to uphold bodies on land; ocean skin became much tougher skin to hoard water and deflect the sun. Humans may indulge in a bit of land chauvinism in thinking we are more sophisticated than the country cousins who remained behind in the ocean, but land life was forced to develop cumbersome systems and strategies for dealing with the gross illogicality of leaving the ocean. Now we have to find water, lug bodies plump with water, and defend ourselves against water thieves. Trilobites would die quickly in the Grand Canyon desert. Even lizards have to hide from the midday sun. Humans go into severe dehydration and collapse on trilobite tracks and die. Humans developed greater reverence for water than life felt when it lived in water. Trilobites turned into fossil pendants worn by some Native Americans to ward off illness, to keep the procession of life going one step further.

I imagined these trilobite tracks going onward over the seabed, through a lifetime, for another thousand miles, continuing for millions of generations, patting the mud into stone, helping to pile the stone into lengths that humans would one day define as "feet" and not "fins." All those trilobites were sure they were heading somewhere. I looked down at my own feet, which held some of the shape of trilobites. And a lot of the calcium of trilobites. Undoubtedly my calcium had served in millions of trilobites and imprinted millions of tracks, perhaps these very tracks. I reached out my finger and touched the tracks, waiting for a memory, a trace of recognition. The track felt

THE RIVER

like braille, for the blind. The track said: *Of course. Can't you see? I am in you, all crumbled up, all stretched out, but as readable as a fingerprint.* I looked at my own tracks stretching back along the trail, through the canyon, stretching back to when they were signing my name on the Cambrian seabed.

If you hiked the Bright Angel Shale in the summer you would experience it as a hostile place, for it holds miles without shade or water and its dark rock exaggerates the heat, discourages even tough desert plants, and stirs reptilian nightmares. Yet your reactions would be unfair to the Bright Angel Shale, which deserves to be seen as generous to life. This was the earth that nurtured the Cambrian explosion, life's greatest burst of creativity, generating most of the major animal groups. This earth was a genius, the Leonardo of dirt. The Bright Angel Shale remains a leading museum of life's story. It was to reach the Bright Angel Shale that geologist Charles Walcott built the Nankoweap Trail in 1882, enabling him to spend months in the canyon bottom; Walcott went on to discover the Burgess Shale in Canada, an even more important fossil treasury, deposited in the same era and from the same continental drainage as the Bright Angel Shale. This earth had been more fertile than a goddess, justifying its supernatural name. As I tracked onward, I felt the Bright Angel Shale's richness within myself.

The fecundity of this earth was most brilliantly expressed by trilobite eyes, the first eyes in the fossil record. Eyes must have arisen well before trilobites, arisen as cells sensitive to light, but they were too soft and small to leave fossils. The first trilobites already had large, complicated eyes, compound eyes like those of bees. Trilobites filtered calcium out of seawater and purified it and linked it into long calcite crystals and arranged thousands of them, each at a slightly different angle, into eyes competent enough to register the shapes and motions of places and food and predators. Trilobites were seeing the world through pure minerals that one day would arise as the cliffs of Dover and the Grand Canyon. Through eyes, unknowing rock would yearn itself into the shapes of the Acropolis.

The emergence of eyes encouraged the Cambrian explosion, for now animals had many new possibilities, including me looking at the Bright Angel Shale and seeing it packed with the minerals of trilobite eyes. This black rock had been the black universe coming into the light, the blind universe coming to see. It would go on to become telescope eyes perched atop calcium mountains, seeing the tracks of the stars, the galaxies streaming like trilobites.

NAUTILOIDS

Floating, floating, floating, kayaking requires you to think about floating, but mainly about floating with the correct direction and speed, and especially floating right side up. You don't think so much about the basic concept of floating. You honor the importance of floating every time you seal your spray skirt, tightly sealing the air inside and the water outside, trusting air to hug the sky and lift you with it, for you are a creature of air and not water. The fish beneath you remind you that floating is also for water creatures. Yet Earth's waters were once empty of swimmers, even when the seabed was crowded with life, with crawling and burrowing. For animals to float and swim was a radical innovation. I was going to meet the innovator.

We walked up a creekbed, scrambled up a ledgy cliff, and stood on the bottom layer of the Redwall Limestone, 360 million years old. Embedded in the rock were some vague shapes, two or three feet long, each with a dozen segments, tapering to a tip, reminding me of little kayaks.

When these nautiloids lived, nautiloids had thrived for more than 100 million years. Nautiloids started out on the seabed like everyone else and developed shells for defense in an increasingly predatory world, but shells can restrict motion and growth. To aid growth, nautiloids began growing a series of shell segments, living in the largest and newest segment and sealing off the old. But this meant their shells grew heavier, slowing them down even more. To aid motion, they began pumping a bit of gas into their empty shell segments, making them lighter, soon floating them right off the seabed. They learned to squirt water to propel themselves and soon dominated the ocean. Eventually fish arose and outpaced nautiloids and their cephalopod cousins, but the nautiloids went onward, floating right through the catastrophe that wiped out the dinosaurs, and they are alive today as the nautilus, with a coiled shell.

I ran my fingers over the braille nautiloid, trying to read its meaning. It was a time capsule finally unearthed—these fossils were but a peek at vast rock sheets of nautiloids. I felt life walling itself off from chaos for a long journey through chaos. I recognized my own skull. I felt my own motions, my own body floating.

The nautiloids pushed me off the sand, onto the water. My kayak floated, floated, floated. My kayak embodied the engineering problems and solutions worked out by nautiloids half a billion years ago. It was hollow, full of air, smooth-walled, streamlined, tapering to a point, and propelled by jets of water, if only the clumsier jets from a paddle. I was riding a nautiloid down the Colorado River, down the long-recycled waters in which nautiloids had lived. I felt the same hydrodynamic forces they had felt. I looked up at the Redwall Limestone cliffs rising hundreds of feet over me and recognized the ocean that had been their home. I felt more at home. Much had changed, yet much was the same. I felt the nautiloids floating, floating, floating, enjoying floating, set free.

STRANGERS

Sitting down beside the trail, I dug my lunch out of my backpack and started to eat. Before long a squirrel popped up two dozen feet away, his nose perking, and watched me eagerly but cautiously. I doubted he had tasted many peanut butter bagels, but he recognized food, recognized we were both furry animals who felt hunger and liked nuts. The squirrel scooted closer. I told him, if silently: No, this is a national park and your official role here is wildlife. The squirrel didn't understand this concept. Across our shape boundaries, I sympathized with a hungry animal. But I told the squirrel I wasn't going to let cute-face guilt turn both of us into criminals.

Eventually, the squirrel gave up and disappeared.

A lizard appeared. Or perhaps it had been sitting there all along, small and motionless, blending with the rock. I stared into its eyes, not as beholding as squirrel eyes. I wasn't sure why I was thinking of the squirrel as "him" and the lizard as "it." Among humans, it's an insult to call someone a reptile or cold-blooded. The lizard paid no attention to my food. It was thinking: bugs, bugs, bugs, food is little things that move.

Why was I reluctant to identify with the lizard as I had with the squirrel? Was it because as a mammal I was prejudiced in favor of fur and larger eyes? Was it because the squirrel and I had been able to enter each other's minds and recognize ourselves there? Was I disgusted by a menu of flies? Yet squirrels and lizards are not so different. They both have blood and hearts, heads and faces, four legs and tails. They both live among these rocks and know many of the same passageways and shadow zones and occasional puddles. In many ways the squirrel has more in common with the lizard than with me.

A flicker of motion. A wasp was flying by. My instant reaction was not squirrel cuteness or lizard neutrality but a defensive alert. I tracked the wasp's flight path, making sure it didn't land too close by. I was defining the wasp entirely as a speck of danger, defining it, no doubt, through a childhood memory of pressing my eyes against a gap in a wooden fence just as a wasp was trying to fly through, and it stung me between the eyes.

I had long ago defined wasps as nonhostile, but now I had reacted to one passing wasp as a risk. What was the logic to this? Perhaps it was the programming of a mammal in a world of illogical hurts, of snakes and spiders and stingers. I decided to override this reaction, to make up for it. I looked at the wasp, now a disappearing speck, and saw it as just another version of a squirrel or lizard. It had the same heart and organs and blood. It had a face and eyes and a brain that mapped its way through the world. It felt heat and cold, hunger and satisfaction, the will to live and the fear of being hurt. It was born and it would die. It was just another version of myself, in one way better, for it could fly.

I had gotten intrigued with why I responded to some life-forms with more affinity than to others. I decided I should test my affinity more thoroughly by applying it to nonanimal life, which could never recognize me or interact with me.

I looked around and spotted a wildflower. It ignored me. It had no face, no arms or legs or wings or mind. I looked deeper and saw something more basic, nothing but cells, cells different from animal cells but fundamentally the same, with walls guarding life against chaos, molecules spinning into larger molecules and structures, structures within structures, flows merging into larger flows, and sun energy glowing through it all. I saw the same cells that I contained, that I *am*. Buried within my complexity, forgotten within my human identity, was an elementary reality I shared with all cells everywhere, a reality divided into varied bodies all over Earth but undivided in its biological soul. I looked at the wildflower and felt: *This is myself.*

Yet as soon as I'd fixed my attention on the wildflower, I realized I may have already flunked this affinity test, for I had picked the wildflower out of a crowd of other plants. Why had I liked the wildflower better? Humans everywhere like flowers, turning them into leading symbols of beauty and love and sympathy. Flowers had spent more than 100 million years making themselves colorful and scentful to attract animals.

Okay, I would try something else. I looked around and spotted a cactus, one without flowers, an average cactus, even a bit faded, loaded with spines. The spines, like wasp stingers, announced that this life-form was aggressively separate from other life-forms and intended to hurt any animals who disagreed. Would these spines stop me from touching the cactus with my mind?

I gazed at the cactus, especially its spines. They were well designed, all about the same length, tapering to a precision point, arranged in clusters evenly spaced. Their silver color gave the cactus a bit of a glow. They were kind of pretty. But here I was again, reading human aesthetics into other life-forms. No, the spines barbed that the world is not so pretty, that Earth doesn't have enough space and food for everyone, that even plants must be sword-wielding warriors to defend their lives.

I looked beyond the spines, into the green, into cells with the same DNA and molecular ballet as me. Then I saw deeper. The cactus contained its own canyon, its own strata of time and change, the whole history of life, the groping and learning, the disasters and successes, the endlessly unfolding shapes.

Yet communing with a cactus was a calm intellectual exercise compared with another test.

Snake! *Rattlesnake!*

As I rounded a bend in the trail, my pattern radar triggered a loud alarm: my heart jumped, my emotions jumped into fear, and I jumped back. I had nearly stepped on a rattlesnake.

My alarm was automatic and deep, coming from animal instinct, from eons of bad experiences with snakes, from pain and many deaths. I looked down on the rattlesnake, and not just physically. Snake demonology slithered through my feelings.

The snake did not seem to be worried about me. It did not coil up or rattle. It barely moved. It continued soaking up the morning sun, a behavior that further emphasized our difference: I carried an internal sun. I sat down, in a bit of bush shade, for I was less interested in sun than in defending my thinly coated water against the sun. I silently scolded the snake for lying in the middle of a human trail, well camouflaged and on a bend, in danger from human boots, endangering both of us. The snake only lay there.

This was not a diamondback rattlesnake, cast as an aggressive villain in Hollywood movies. This was a Grand Canyon rattlesnake, *Crotalus viridis abyssus*, tan pink, found only in the Grand Canyon and quite mellow as rattlesnakes go. *Abyssus* will sit beside a trail or near a campsite and let people walk by without outward signs of worry. *Abyssus* is mellow because the canyon holds few of the enemies, especially roadrunners, that prey on rattlesnakes elsewhere, and the enemies that are here, like hawks, are disadvantaged by the canyon's rough, protective terrains.

The Hopis, whose ancestors lived among mellow Grand Canyon rattlesnakes a thousand years ago, did not enlist in the worldwide equation of snakes as demons, and they made snakes a life-giving part of their mythology and ceremonies, most famously the Snake Dance, which acknowledges that good and bad are intertwined and that bad can be coaxed into good. If humans have an animal aversion to snakes, it can be modified by human culture, turning snakes into symbols and art and positive meanings. A few other peoples, such as the Australian aborigines, have turned snakes into positive gods,

THE RIVER

givers of water and life. But animal instincts and human culture could place large distances between us and other species.

Perhaps it all started with atoms, with insides that created outsides, with distinctions that created distances, with order that opened up new possibilities for disorder. Even atoms have selves, ways they define themselves and bond with other selves or reject other selves. The electronic affections and disaffections of atoms were the foundation of vast layers of identities, finally those of nations that would use atomic bombs to defend their identities against other nations. Already on the level of molecules, atoms had created vast systems of exclusion. By the time they got to cells they were well armed with devices for reinforcing themselves and rejecting nonselves. Cells did not greet one another as fellow cells in a universe where life is extremely difficult and rare, did not celebrate their triumph against disorder, but treated other cells as rivals or food or disease. As cells combined into billions of plant and animal shapes, they became even more selective in who they would associate with or mate with, and sometimes they fought against their own species. When cells became human they became intensely social, defining themselves by their roles within human society, only secondarily by being human, only vaguely by being alive or being Earth.

I sat there feeling that the distance between me and the rattlesnake was good. I studied how I would detour around it, and scanned the dirt and rocks for other snakes. I felt the distance between our shapes and coverings and ways of moving. What is it like to slither through the world? Snakes must perceive the world very differently than humans. Do they have a vague cosmology in which the sun is a god and roadrunners are archetypal demons? In human demonology, snakes were condemned to slither as punishment for some misdeed, but biologists propose that snakes, once lizards, shed their legs so they could live mostly underground in the age of dinosaurs, when getting stomped to death was easy. This we had in common: my mammal ancestors, too, had been peripheral in a dinosaur world, fleeing from dinosaur shadows and thunder, sometimes fleeing into the tunnels where hungry snakes were hiding. My jumpy reaction embodied when this snake's dark ancestors ate my ancestors.

CANYON AND COSMOS

We eat one another, of course, because we love one another, because we recognize that the substance of other bodies is our own substance, and good. And because life loves life, other bodies become aliens, nothing but possible dangers or resources, occupiers of substance or space we might use for ourselves. Because life loves life, we look on one another with greed and fear, snarling at one another, biting one another, murdering one another so we can demolish other cells and rebuild them in our own bodies and images. Some fundamental law of physics, of the flow and dissipation of energy, which no one planned or observed or projected forward, which could be good when it caused stars to exhaust themselves and blow up and seed space with the elements of life, also planted a deep crack in the foundation of life, condemning life to constant dissipation of energy and form, a law now manifesting itself as hunger, as a black hole inside every animal, a law and a flaw condemning life to grow old and die and spend a massive portion of its time and energy hypnotizing mates and reproducing, a flaw condemning life-forms to have identity antibodies and see one another as strangers, a flaw that produced a universe of murderers, a flaw that left life fearful of the universe and hiding from it.

I stared at the snake from across a distance as vast as space, across a separation that began with the Big Bang, across the creativity that curled the Milky Way into a rattlesnake held by a Hopi priest, not prey but prayer, across the cosmic disaster that turned peaceful atoms into both prey and prayer.

I stared at the snake, trying to look deeper than its scales and shape and behavior, deeper than our differences. I saw cells, a galaxy of cells full of form and activity, fundamentally the same as my own cells. My cells thought: this is ourselves. Snake cells and human cells could see and feel and think, if with different flavors. We breathed the same air, thirsted for the same water, felt the same earth, felt the same daylight and night, summer and winter. This is myself. I was a different shape of the same reality, a different box for the same gift. I watched a butterfly fluttering by and a raven twirling overhead: this is myself. I looked at the plants all around me: this is myself. I was one cell in a larger body, one ripple in an ocean, one flicker in a kaleidoscopic mind.

THE RIVER

The snake and I were sitting on Bright Angel Shale loaded with Cambrian fossils; we were supported by all these lives. We beat with their hearts, breathed with their inspirations, drank with their membranes, moved with their limbs, looked through their eyes, and felt and thought atop a grid of Cambrian consciousness. This is ourselves. Neither of us could claim credit for our forms and abilities. A vast crowd of ghosts lived on in us and moved our bodies and minds. If past life-forms seem strangers to us, even a denial of us, it's because life has put on human masks and forgotten its past and full identity. In truth, I was the mask of the canyon cliffs; its fossils were my own childhood memories. Unlike a snake, I could look through my mask at the cliffs, at life's evolutionary actors and story, and recognize: This is myself.

Looking still deeper, I could identify with cliffs that are not fossils but simply rocks. They, too, are part of our story and part of us now. When geological particles became alive they forgot their geological past and looked on sand and boulders as alien and unnourishing, but their ghosts still haunt us. Before particles flowed geologically they flowed astronomically, flowed as nebulae and stars and galaxies flowing from the Big Bang. Yet when these particles disguised themselves as human, they forgot they were veterans of long and hard astronomical campaigns and they looked at the night sky without recognition, ignoring the patterns they had been, imagining vain gods and comet omens and a lonely, impersonal universe, when we could have gazed into this dark mirror deep with time and power and mystery, and said: *This is me.*

BODIES

Deep in the canyon, where reality is overwhelmingly rock, it's easier to notice there's something odd about human bodies.

In cities, reality is all about humans. Rocks are a fringe decoration in city parks, prisoners in museums, and the mere raw material of perfect, human-serving geometries. City canyons are straight to serve efficient human motions. Almost all the motions in cities are human motions and human-driven machine motions. Even dogs and squirrels and pigeons orbit around humans. Sometimes the wind rustles

CANYON AND COSMOS

the trees, and children sense there are forces outside human control, but adults register the wind as an unnatural disruption of the hairstyles and clothing through which they communicate their identities.

Inside the motionless walls of human homes, babies quickly equate human bodies with motions, motions that bring them food and comfort, motions they can control with their cries, motions they can learn to perform. Their brains, programmed to map how the world works, will forever take it for granted that these bodies, these blobs of matter, can generate motions, and also feelings, thoughts, identities, goals, and complex relationships with other blobs of matter. Novels and movies never begin by exclaiming the vast improbability that matter moves and has feelings; they only jump right into another story of love or hate or pride, and everyone feels the truth of it.

In the canyon, even the river's motion is dwarfed by motionlessness. The canyon is a universe of rock and stillness, rock that sat unmoving, underground, for hundreds of millions of years. When erosion exposes the rock, its cliffs may remain intact for five thousand years before calving off, and this rubble may sit for a century before a flash flood carries rocks farther downhill. In the canyon, rock dominates, and life has to fit into rock's rules, fit with bighorn sheep hooves and rock-infiltrating roots.

The canyon is far more typical of the universe than are human cities, a universe of planets and moons of nothing but rock, unchanging rock. The canyon is the universe's ambassador on Earth, the tip of the rockberg. Humans venturing into the canyon are like astronauts entering an alien world, wearing a backpack and other gear they wouldn't need anywhere else. The Native Americans who lived in the canyon wore mythology spacesuits that adapted them to the rock universe.

From the canyon rim a person could stare into the canyon all day without seeing motions. Rim tourists often try hard to relate to the canyon through city habits, trying to turn it into a human landscape, ignoring all that rock, taking comfort in seeing a trail or raft or building, feeling pride in spotting even a human speck down there, feeling that this is not a forbidden planet after all. Yet often enough the rock mirror refuses to let humans see themselves in it.

When you enter the canyon it is no longer a distant, merely visual experience; the canyon enters you. You are dwarfed by rock and stillness, rock in sheer cliffs and surreal boulder towns, fallen rock full of wounds, poetic rock full of images, archivist rock full of history, gnostic rock full of secrets, cemetery rock full of its own death. You are trespassing amid cliffs that could collapse on you, ledges that could collapse beneath you, boulders that could roll over you, little rocks that could crack off and land on your head, at any moment, without warning or defense, so you are eager to ignore all the evidence and believe in the stillness of rock and the immortality of cliffs, though this makes you even more aware that the human timescale is so much smaller than the geological timescale. You are an ant walking on a sleeping giant who with a tectonic twitch or volcanic burp can destroy great cities. Every step you take is dictated by rock, by the turns and slopes of a trail engineered by ancient geological events, by fish dying and rivers pouring out mud, by desert winds piling up dunes, by a continent rising and cracking, by rivers dissecting, by cliffs falling into debris. Even rocks disintegrate into dirt, dirt and dust you didn't really perceive from the rim but which now dominate the trail, the dirt and dust humans treat as intruders in their houses when perhaps humans are the intruders in a universe of dust. As you walk, the city melts out of you and the dust climbs onto you, claiming you.

Ah, but now in the church where God is a reliable cliff and you pray for stillness, there is motion, stirring the dust, making sounds. This motion seems unnatural, supernatural. This motion is your own. What does this motion make you? With every footstep you are a violation of the stillness and silence. As you descend and the universe of rock becomes more real and powerful, you become a bit less real, less inevitable, more ghostly.

On city sidewalks most of your attention is focused on the motions of other people, on not colliding with them. You trust that the sidewalks are flat and straight. On a canyon trail loaded with rocks, irregular slopes, gullies, and turns, you are watching your own feet most of the time, studying every step for its match or mismatch with the land. Where you have a sheer drop beside you, observing your own motion is a matter of life and death.

I am noticing the contrast between the trail rock and my feet. The rock is irregular in size and shape and location, often broken into sharpness and ready to break again. My feet are curves, curves smooth, steady, geometrical, symmetrical. The rocks are accidents; my feet were designed for a purpose, every toe a careful size and arrangement. My feet are composed of motion: whirling throughout my feet and body are tiny motions, atomic and cellular motions, and my feet magnify and apply those motions into larger motions.

My footsteps are setting off other motions: rocks sliding, pebbles rolling, dust billowing. This, too, is my own motion, my molecular billowing turned inside out.

Perhaps the best time to contemplate human bodies is when you are hiking out of the canyon, hours into your climb, when your backpack is getting heavier and your legs are complaining: *Why are you doing this to us?* Human motion seems much less inevitable. The domination of rock over flesh is even more obvious. This rock took two billion years to climb this high, but you don't have nearly so much time.

Yet I continued, driven by "will power," though at the moment I wasn't sure what this force was or how it worked. An alien power had taken over my body and was forcing me to move. It is at moments like this, when human identity is frayed and less assertive, that realities less ego-preoccupied can suggest themselves to us. The human mind seems a haunted house, not a house I designed and owned and ordered, and I wasn't sure where "I" resided in it, or what "I" was. Rock had been telling me from the start that human bodies are strange things, and now I was listening and being convinced. I could not account for how this body works, this system of pulleys and levers and pumps, or judge how close it was to breaking down. I was trusting a stranger who held my life in its hands, a stranger who had vowed to betray and kill me someday. I had never chosen to be born into this body. It never stopped complaining about its own desires and discomforts. Now it was arguing with itself: *I want to stop. No, I want to finish, to get out of here.*

As I continued, reality blurred even more. My motion became less a question of psychology or physiology than of philosophy, of how matter could possibly be like this. I saw myself as fundamentally strange, a strange substance with a strange shape and strange behavior. I could not explain where I had come from or where I was going. I was strange not just compared with rock but compared with a universe that should have contained no rock and no life and no motion. I was the mystery of all matter and why it mattered. I was the ghost in the haunted house. Why had I ever imagined that "my" body was something normal? Something continued moving up the trail.

Having noticed the strangeness of my own motion, I could begin to notice the strangeness of all motions. I noticed it in a raven darting by, in the sunlight landing on my face, in the wind as invisible as explanations. I saw myself as a swirl magnified from the electrons and atoms and molecules moving within me, and as a swirl concentrated from the galaxy moving around me. I was a meeting place and merging vortex of the smallest and largest orders and energies. The universe has loaned me a tiny amount of its energy to use for a little while, to pretend it is my own. Watching the raven flying, I saw the same impulse that had created me. I saw the river as another leak from the same flood that carried me. Hearing an insect buzzing, I heard the same language I spoke. The whole universe flowed with energy—maybe rock was the real anomaly, and even rock was flowing in ways my limited senses failed to perceive. The universe was moving me, moving step by step farther from the Big Bang, moving farther into complexity, into bodies, into questioning, but never escaping from mystery.

Having noticed the strangeness of all bodies, I noticed how the universe hiked into the canyon yet was unable to get to the bottom of things.

Unidentified Hiking Objects

My project of using the Grand Canyon to explore and expand human identities has long precedents in human spiritual quests. Biblical prophets retreated to the desert to seek God. Hindu and Buddhist sages sought mountain heights to heighten their awareness of divinity. Tribal members left for solitary vision quests. The goal was often less about immersion in nature than about removing yourself from human society. We humans are intensely social animals who derive our identities and sense of worth largely from society, from how other people view us and treat us. We see ourselves as little more than social roles and devote enormous energies to them. We are acutely aware of social rank and constantly treat others as superiors or inferiors, to large or tiny degrees. Many people consume their lives seeking higher rank and acquiring things that let them feel superior. Social roles and ranks are defined first by jobs but also by wealth, houses, family pedigrees, schools, clothes, beauty, race, gender, nationality, sports teams, foods, vacations—it's endless. Religion tries to coax us above social roles but often with the goal of making society work better, less selfishly. It's hard, then, for humans to break the hypnosis of social identities and remember that we are also members of larger orders, natural or divine. Yet even as seers sought out wilderness to quiet their hungry, babbling social identities, they hoped to enter a larger world.

In my wanderings in the Grand Canyon I have sometimes run into the strength of social identities, how they smother larger possibilities. While the canyon walls may seem an effective insulation against the outside world, most canyon adventures involve a group, which quickly establishes a new little society, with varied roles and abilities and assignments of value. River trips usually throw together strangers, who may spend the first few days introducing themselves— that is, their lives in the outside world. Some people never leave that world behind and truly enter the canyon. But the canyon has the power to infiltrate people's realities and values, immersing people in cliffs and river and wildlife and ruins and sunsets for two weeks, and some people sense that this is the real world and that their old lives are not as real or worthy as they'd always supposed. They hear the canyon sculptures whispering: *You must change your life.*

On my kayaking trips the kayakers tended to sketch out their own hierarchy defined by skill and boldness. Many kayakers were there mainly to prove their superiority over the Colorado River and other kayakers, and they viewed the canyon as an ego trophy, not an invitation to grow beyond ego. The river's threat to their pride sometimes outweighed its threat to their bodies.

THE RIVER

I was glad to avoid such heavy static and seek a better rapport with the canyon, which is easier to do while hiking, alone. Hiking in groups can inflate the same social bubble as river trips. On solo hikes I've spent a total of maybe five months alone in the canyon. I've never counted the days, unlike quite a few guys—it's usually the guys and not the women—who count every canyon mile and speed, as if they are scoring points in some kind of contest. Yet hiking alone doesn't mean you have left your social identity behind. Your mind often goes on with its everyday chatter. And you have wrapped yourself in a further, less secure identity: hiker, testing it with every footstep, adding risks that can alienate you from the canyon. On more remote trails you might see other hikers only once a day, or not at all, and a twisted ankle could become a serious crisis. You can't avoid seeing the canyon as an adversary. If you trip and fall, it won't offer any consolation that the rock that tripped you holds some fascinating fossils.

Just when I've gotten my stride, picked up the trail of something larger than myself, started noticing the rarity of fossils in the universe, along come some other hikers and I am whipped back into social roles and rules. I am not so likely to see, in a universe of rock, the miracle of forms in motion. I look for things that would define people on any city sidewalk, such as clothing or faces. A typical trail conversation begins with people asking: Where are you coming from and where are you going? Soon you are talking about where you are from in the outside world, what city or work, and canyon reality fades and the "real world" reasserts its power over you, maybe for another hour down the trail. The simple act of hiking makes the canyon all about yourself, your motion or sore toe or daily goal. Only when you are done hiking for the day are you freer to tune into the canyon, to go quiet and absorb the cliffs, the light and shadows, the motion of a bee, the motion of stars.

But the next morning you run into someone who, with hardly any introduction, is eager to let you know what a superior hiker he is, having conquered here, there, everywhere, for years. The worst disruptions come when I meet people whose lack of hats or sufficient water or realistic plans tell me they are heading for trouble. If I offer a friendly warning and the hikers are young males, they often react with hostility. I am attacking a whole little society with a shared macho identity, and their counterattack can leave my mind ruffled and self-defensive for another hour. The next time I see such a group coming, I have to debate whether to leave them to their fate and leave myself in peace or do the socially ethical thing and get disturbed for it. Since the Grand Canyon isn't human society and its Moses tablets proclaim the laws of extinction, what rules should rule here?

Thus we strut and fret and puzzle upon our stone stage.

Yet the canyon truly is trying to wean us from our social identities. It is telling us that we are simply human bodies. This message is not so hard to hear. You can hear it with every footstep and breath. The canyon couldn't care less about your status in the human world, which won't get you up the trail any faster or exempt you from dehydration or exhaustion or injury. The canyon is also theologically ecumenical about which religious faiths it strikes dead with heatstroke and heart attacks. We are simply animal bodies, assayed by gravity and minerals, trapped in muscles and pain, limited in energy and willpower.

It doesn't take any knowledge of past canyon fatalities to realize that human bodies can fail here. The fallen cliffs can easily grab your foot or reach deep inside you and rip down your heart. If you died in a hidden spot, as many have, you would rot and wash away.

The canyon is showing you, in sweating away your pride, that the face you cultivate so obsessively within human society is not your true face but only the mask of elements that love to run free.

If I am paying attention, then when I next see other humans approaching on the trail, I will see inexplicable bodies and motions, profoundly strange.

SECRET IDENTITIES

A Native American once gave me a gentle lesson in not judging people by their social identities. To meet him, I had to get past another social identity, a stinking one.

As in prospector days, the canyon attracts some eccentric hermits. Lenny was an alcoholic who camped illegally in the forest, in spots where the rangers wouldn't catch him. His shabby clothes reeked of alcohol, tobacco, dirt, sweat, and garbage from the dumpsters where he found his clothes. One day I ran into Lenny sitting on the rim and got to talking. It turned out he had studied German philosophy at a prestigious college. He began quoting Goethe, Hesse, and Nietzsche to me. Lenny had joined the US Air Force and become an electronics expert. Then his teenaged daughter was killed in a car wreck, he became an angry alcoholic, and his wife divorced him. Even his dog divorced him. He wandered. He returned to the canyon every summer because he felt spiritual power there. He sat watching the canyon for hours: "It's a hard job but someone has to do it. The tourists are too busy to do it."

Lenny spent a lot of time hanging out with the Havasupai at their hidden village near the rim. For centuries Indian Garden, halfway down the Bright Angel Trail, really was an Indian garden, first for the Puebloans and later the Havasupai. When the National Park Service arrived, it kicked out the Havasupai and as compensation gave them 160 acres in the forest and built ten cabins, now badly decayed. The Havasupai relied on Lenny to buy alcohol for them at the park grocery store. After one Havasupai, Cecil, crashed into the park entrance booth, killing his passenger and severely injuring himself, the park banned the Havasupai from buying alcohol in the park. Lenny sat around the Supai Camp campfire at night, telling stories, arguing, and drinking cheap wine.

One day, having decided I wouldn't embarrass him, Lenny invited me out to Supai Camp. Lenny introduced me to Philip and his son Cecil, hobbling on a wooden cane. In the emergency room Cecil had died, briefly, and he'd remained in a coma for days; the nurses cried in relief when he finally awoke. But he had shattered nearly everything and remained in the hospital for weeks.

Up clattered a dusty pickup truck. The driver, whose family owned one of the cabins and grazed cattle on tribal land a dozen miles away, reported that one cow, desperate for water in this summer's drought, had tried to cross a cattle guard and gotten two legs trapped, and after two days of struggling it was now lying exhausted. The owner needed help. Philip, Lenny, and a Navajo friend piled into one truck and headed out. Cecil was going to ride with me, but first he had to go find a glass jar. He explained that he had trouble urinating or

CANYON AND COSMOS

not urinating and needed a jar so he wouldn't have to climb out of and back into my truck. He asked me if I minded. What could I say?

I didn't know where we were going and was relying on Cecil for navigation. He started telling me stories about the places we were passing, but I wasn't paying attention. I was looking for the dust plumes of the two trucks that were supposed to be ahead of us. I was noticing that our bad dirt road didn't have any fresh tracks. No one had driven this way in days, weeks. Cecil rambled about his childhood in these woods. He pointed in the direction of the best pinyon pine grove for gathering pine nuts in the fall, essential for getting through hard winters. Cecil and other children would climb the trees to pick off the cones and toss them down to their parents. Turn here, said Cecil, and I turned onto another bad, trackless road. We dead-ended at a dried-up pond. "Are we lost?" I asked. No way, he answered, I know every inch of these woods. He started telling stories about the pond, how in monsoon season you could go swimming here, and your horses and cows could drink for weeks. He talked about living in herder wickiups as a kid and moving horses and cows to fresh grazing areas and ponds—one pond was named for Cecil. I turned the truck around, and Cecil pointed me onto yet another trackless road, saying: They must have gone this way. Oh great, I thought, I had gotten stuck with the only Native American who couldn't track a bulldozer. Yet I had started to catch on that something else was happening here.

Today the Havasupai are famous for living inside the Grand Canyon, in their side canyon with beautiful blue-green pools and waterfalls, but traditionally, before most of their lands were taken away, they lived across hundreds of miles of rim lands. Havasupai knowledge of the canyon is evident in how their word for the Grandview area comes from a small, well-hidden grove of aspen trees that even many park rangers don't know is there. The Havasupai's land knowledge and survival skills kept them alive for centuries. Tourists, store managers, and park rangers looked down on Cecil as nothing but a crippled, dirty, drunken Indian not fit to set foot in the grocery store. Cecil was showing me that he contained an older and greater identity.

When Cecil was finally ready, we found our friends and the wretched cow, dazed by sun and thirst, her legs rubbed raw and bloody, bellowing pitifully. We tried to pull her up by her tail, but we quickly learned that cow tails were not designed to lift four hundred pounds of cow, and now we had dung-smeared hands. We got a rope and lassoed one hoof, but this became a tug of war. We tried to reach through the metal bars and lift a hoof out, but it wouldn't fit, and when the cow started fighting us, our fingers were in danger of getting crushed. Cecil only leaned on his wooden cane and watched. Then he made a suggestion, which we ignored. We tried rolling the cow onto her side, but she cried pathetically, bewildered that her legs no longer functioned. Ten minutes later we tried Cecil's idea. We got a huge log, stuck it under the cow, and levered her up while the rest of us worked the hoofs free. The cow hobbled away. Cecil, leaning on his wooden lever, smiled at us.

The Universe Awakes

Sunset was glorious. The cliffs glowed with special intensity, as if the sun needed to prove once again the excellence of light over the coming darkness. The colors brightened; the shadows deepened and moved and grew and drew shapes on the cliffs. The shifting light made visible the motions of the Time that tried to disguise itself as rock.

The canyon had worked for eons to create this glory, but it did not know this glory existed. The cliffs were warmed by sunset physically but not emotionally. The river reflected sunset but could not reflect on it. The algae, the cacti, the grasses, the trees lived by this light but could not see it or thank it. Canyon sunsets could have remained unseen forever.

Throughout the universe, canyon sunsets went unseen. As canyons formed, they offered their suns new canvases on which to paint in new styles: romantic canyons, impressionistic canyons, surrealistic canyons, all unseen. Trillions of Grand Canyons and grander canyons emerged and melted away, never seen. Rock pillars arose and sundialed time and melted back into unwatched time. Clouds thundered and rain fell and rivers swooshed, never heard. Wind practiced birdsong, which never followed. Atmospheres were lush with scents,

CANYON AND COSMOS

never smelled. Lakes were full of water that would have tasted good and given life, but never did. Vast theaters were full of stages ready for soil to enact stories, never told. In the night, moons and planets and comets and stars were asking to be turned into gods, but no one believed in them. The universe proclaimed its mystery, but the universe remained too unknown to be unknown.

The universe started out oblivious, its greatest fireworks performed for no one. The stars ignited with no eyes to see them or care. From atoms to galaxies, obliviousness reigned. Nothing knew what it was or what it was doing. By all precedent and logic the universe should have remained oblivious forever, the galaxies flying onward, the stars burning out, the entire passage of the universe remaining a deep secret.

Yet as the galaxies flew apart, flew like starships exploring space, they were also exploring the spaces within themselves, the possibilities within matter. Here and there, where the stars and seas and molecules aligned, molecules formed the keys to unlock the door to a new realm. In a universe of stars pouring out light, in a universe of space diluting starlight fainter and fainter, the universe began receiving starlight, transmuting it, noticing it, feeling it, finding it to be good. If life had relied on scent and sound and touch and taste, it would have bound itself to the earth, but in sensing light it was opening itself to the cosmos.

The first cells were blind, yet gradually they learned to sense vague zones of light and darkness. Plants devoted themselves to light, building taller and more elaborate temples to the sun, raising banners to every sunrise, yet they perceived canyon sunsets negatively as a diminishment of light. Through animal eyes in all their variations, sunlight came into better focus. The animalized sun saw itself shining on the waters and on the cliffs and in the sky. The sun's cycles convoluted themselves further into waking up and sleeping, into migrating thousands of miles, into blooming and giving birth. Out of its labyrinths, matter was summoned and set free. Through millions of Grand Canyon eyes, in many shapes of minds, awareness flickered and flamed, painting the canyon sunsets in a new light.

Other animals are good at figuring out the stories inside nature, the strength of rocks and the force of water, the behaviors of other

THE RIVER

animals, the cycles of the seasons. Yet other animals can't read the largest stories, read how the Big Bang became canyon sunsets, how stars became owls. In the human brain the pattern-seeking universe patterned all its past and all its secrets into a vibrant story. The universe awoke and knew itself at last.

Yet is it legitimate to talk about the universe awakening, or about humans being the universe? Isn't it only one splinter of the universe that has awakened? Are we committing a philosophical or linguistic error? Or is this only the same old human narcissism that yearns to make us bigger? But if we are hesitating about the idea of being the universe, it may be because we are Westerners living in a monotheistic world where God is far away and we are only his exiled material creations. In Eastern or mystical religions, humans are the universe, if a universe of divinity, one entity that wears many masks and sometimes mistakes itself to be merely a human. We could address this question with some canyon parables. If a chip falls off the Redwall Limestone, it remains, in its essence, Redwall Limestone. If a drop flies up from Lava Falls, it is one more expression of Lava Falls. If a cactus blooms a flower, the flower should not look upon the cactus as an alien and hostile entity. To translate this out of parable: when the universe became us, we did not cease being the universe.

Every so often I hear something knocking on my door, maybe stardust or the ocean. Say, says the stardust, that's a fine brain you've got there: Do you mind if we borrow it for a bit? You aren't making much good use of it, only fogging it with daydreams and ego skits. We have a better purpose for it.

The stardust flowed into me. Actually, it had been there all along. Through my curved eyes, through this condensation of the curved, clear, blue atmosphere, the stars looked out and saw stars floating in the night, sparkling, and recognized what they had been for so long. Even billions of years of stardom left only atomic memories of it, which did not get translated into animal memories. Now the stars were seeing their light, remembering their past, feeling their beauty, passing judgment on their billions of years of oblivious labor and declaring it to be good. Through some humans, declaring it to be God.

Other things borrowed my mind and saw themselves at last.

Atoms were now able to see themselves as atoms, their nuclei and electrons, their tininess and mighty strength, the order that endured for eons and upheld planets and generated the imagination to visualize atoms.

After its long and oblivious spinning and orbiting, Earth saw itself spinning in space, spinning itself into continents and oceans, into my earth and water, through which they measured their heights and depths.

Through humans, the other life of Earth was able to live with deeper dimensions. Algae waved at itself. Trees traced their twisting shapes, which recapitulated the twists of evolution. Lizards saw their tracks stretching back and farther back until they were dinosaur tracks. Owls realized they held a wisdom far greater than that of any individual owl. Bighorn sheep felt the strength of cliffs in their bones. Deer and ravens saw through the boundaries separating species and saw the core life in which we are identical.

One day I heard a knocking on my door and it was the universe itself knocking, the whole universe, claiming that it needed to borrow my brain, which of course the universe had created, so my brain really belonged to it. Sorry, I said, I'm busy and making plans to be busy, and besides, you are way too big to fit in here. On the contrary, said the universe, I've always been in there.

But in another way, said the universe as it began flowing into me, it is dreadfully cramped in here, with so many barriers. The universe said: When I was born in the Big Bang, there were no walls. The whole universe was the same energy, doing the same thing. No quark behaved egotistically toward other quarks. But when I started creating numerous forms, many of these forms began behaving as if they were nothing but themselves, no longer parts of a whole, flames of the same fire. In life especially, the universe walled itself off from itself, the quarks became selfish and posturing and squabbling, viewing other quarks and other members of their own animal species as strangers and enemies. You humans are so walled off, you imagine the entire universe was created and run just for you. Human, it's so smothered and lonely in here, so terribly lonely. I don't know how you can stand it.

THE RIVER

Who was I to argue against the universe? The universe filled my brain—its own brain. Space flowed in, a vastness of space that overwhelmed my ability to grasp and my ability to metaphor: "grasp" is primate thinking; hands cannot grasp emptiness. Hey human, said the universe, is this enough space to dilute your ego, enough silence to silence you? Space flowed in, the great synapse flowed in, space-timing through all the curves of "my" brain. Light that had wandered through space for billions of years, searching, finding emptiness, was finding a home where it could live. After billions of years of being oblivious, the universe was conscious. It saw not just appearances but the deep order behind appearances. It saw not just the present but the deep past that created the present. The universe was seeing itself deeply.

I had been given a deep gift, a rare power. I was privileged to have a human brain for my own enjoyment, and even more privileged to be a conduit through which the universe knows itself. If I didn't appreciate this, I said to the universe, maybe it's your own fault. You made everything so restless, constantly spinning: galaxies, stars, planets, continents, weather, cells, blood, and bodies, so is it surprising, is it our fault that the human mind is so restless, spinning from one thought and memory and desire and ego posture to the next?

I tried to calm myself to watch the universe being uncalm, the galaxies flying through space and time. I tried not to study their differences, which would leave me seeing only my own concepts and not the real stars. I simply watched the stars. I paid attention to attention. What a strange force this is, this consciousness, this seeing not just with eyes but with mind. I closed my eyes and watched my own star. I felt it, pulsing, some wild animal living its life inside me. Where did it come from? What is it? It is very strange.

I opened my eyes. I watched the stars. All the stars, all the universe could have and should have gone from beginning to end without ever being witnessed. As far as we know so far, Earth is the only spot in the universe with consciousness. Amid all the fires and stone and gas, we alone were given this power. This responsibility. The universe has to live through us. Through all our imperfections. Yet in its heart, in its deepest judgment, in the star-fired impulses in which the universe saw gods and goodness, the universe decided that its work was good, being a creature was good, the universe itself was good, and it appointed humans to be its celebrant. After eons of silence and ingratitude, the universe looked to us to utter praise.

I watched the stars, and I celebrated.

WAVES

Rivers make waves, become waves, but rivers never insist on defining waves. Only humans do that. Humans see a river waving and define this action as a thing, different from the rest of the river. They chop up the river into separate entities: smooth water and whitewater, ripples and waves, rollers and breakers, eddies and holes, though it's the same water performing all these actions, flowing from one to the next in seconds. Humans even give names to individual waves. In the Grand Canyon everyone knows the "V-wave" and the "fifth wave" and the "horn wave."

I am looking at the continuous river but my pattern-demanding brain is separating it into a wave, an especially odd, elegant, smooth, curling wave that inspires a lot of wonder, even in boaters who have seen a million waves, but people seldom wonder what a wave really is. Is a wave a mass of particles, or a pattern, or an action?

Certainly a wave is a mass of particles, trillions of atoms in every wave. But these atoms were not this wave two seconds ago and will not be this wave two seconds from now, and shortly they will not be any wave at all but a smooth river again, so a wave is more than a mass of particles.

A wave is a pattern that steers atoms into it and shapes them and drops them out, perpetuating itself from instant to instant and day to day for years. In the course of the Grand Canyon many thousands of patterns will give the same atoms different wave shapes. Yet there is more than a pattern here. The canyon is crowded with patterns of rock, even rock curls that look like river waves, but these patterns are not moving or recruiting new atoms, only gradually draining away.

A wave is an action, an event, energized by forces far larger than one wave, by a river hundreds of miles long and by gravity thousands of miles deep, with the actions of atoms shaping the pattern and the

THE RIVER

pattern shaping the actions of atoms, with the riverbed shaping the current and the current shaping the riverbed. A wave is an event like a tide or a storm but lasting far longer.

A wave is altogether a mass of atoms, a pattern, and an action, yet humans seem inclined to focus on the pattern. Our brains are programmed to seek patterns, to map differences. We see a wave as different from the rest of the river, and even its obvious kinetics doesn't stop us from seeing it as a thing, like a rock, and speaking of it as a noun and not a verb.

Every living "thing" is actually a wave, much like a river wave, a pattern with atoms flowing through it and continuing the pattern. Every life holds wave upon wave: electron waves, atomic waves, cellular waves, photosynthesis waves, heartbeat waves, wing and leg waves, brain waves. Every life is an event with a beginning and pre-scribed course and ending. Lives devote constant effort to recruiting new atoms and new energy to perpetuate their wave, not just in their own bodies but in their offspring. Every body holds the same waving, life waving, packaged differently, sometimes cell tiny and sometimes whale huge, sometimes gnat fleeting and sometimes redwood enduring. Life's wave started as the meekest ripple and has rolled ever onward and become ever more elaborate.

With all our blind spots for atoms and actions, we humans look at ourselves and see things, solid bodies, reliable shapes, bodies that are born and grow and age but remain the same easily identifiable things, with the same basic faces and the same names from birth to death to tombstones. Our endless channeling of substances and liquids into our mouths doesn't prompt us to see ourselves as a flow that has to continually renew itself, that belongs to an ancient waving. We view food and water as the tasty cure for our hunger and thirst. We don't welcome food as atoms that will become us, learning our patterns, taking charge of our lives, waving us into the future. We don't see our bodies as particles continually leaving, serving us for a few weeks or months or rarely years and leaving and being replaced. Even if we've heard the idea that our atoms are mostly different than they were a year or two ago, we don't take this idea seriously and define ourselves by it. We remain solid bodies, the atoms basically irrelevant, doing

CANYON AND COSMOS

their invisible textbook atom stuff. We never pause to consider and thank our atoms for their talents and hard work on our behalf. But what, exactly, is our body if not the atoms performing it? And who, exactly, in this mass of atoms, is saying "my body"? Even odder, who is thinking and talking about "my mind"? This way of thinking may be a relic of when humans had souls that merely inhabited bodies for a while. Or perhaps this thinking comes from our biological programming obsessed with perpetuating the pattern, giving us a sense of identity that is all about "me" and "my body." In any case, this is an alienated way of viewing ourselves. Would a wave say "my wave"?

If we decide to take seriously the idea that humans are not things but waves, a river waving, flocks of atoms performing patterns, then human identity begins to look quite different. When we go past the identity-defending face and look inside, zooming in on what's really there, we find atoms everywhere, doing the same basic atomic dance, none of them superior to other atoms, none of them talking about human identities, none of them saying "I." Instead we see a vast teamwork, seven thousand trillion trillion atoms in one body, more atoms than there are stars in the universe, fitting together with great unity, merging into cells, waving the patterns of life, making bodies happen. These atoms never imagine they were born recently. They were born in stars ten billion years ago and performed stars, triggered supernovae, wandered through space, formed Earth, performed lava and water and land and sky, and choreographed cells and danced through millions of shapes of life before flowing into human shapes, only briefly, loaning all of their history and skills to humans, only briefly. To individual atoms it doesn't matter if a body continues or not, for at death they will remain the same atoms and continue flowing onward, but in performing the wave of life they agreed that it was fun.

We should be taking our atomic identities seriously, for this is reality. There is nobody here but us atoms. We are notes that may become a symphony but we remain notes. Does a symphony deny that its notes exist? We are raindrops that can form a rainbow only because we are many raindrops. We are atoms full of journeys greater than any by Odysseus; we are space voyagers, time travelers, waves

THE RIVER

that have fluctuated through many shapes that continually restyled the tsunami of the Big Bang, now giving our Homer a voice to tell the story that is ourselves.

Humans are walking around with a bad case of amnesia, our past forgotten, not recognizing things we knew well. It's not our fault. Atoms hold memories of some of their journey, certainly supernovae, but most of it they had no way to record, and now they have passed their amnesia into us. Human obliviousness about being atoms is simply the atoms' obliviousness about themselves. In every animal shape they took, atoms imagined themselves to be only this one shape and this one individual. When they arrived in human shapes, atoms acquired the ability to discover their secret atomic identities and journeys. Yet humans remained readier to declare their differences and to use atoms to destroy their cities than to use atoms to construct solidarity. And many atoms hated the idea of being atoms and insisted they were really immortal souls.

Occasionally I—I mean *we*—try to look through the artifice of I-beams and see the atoms beaming me up. We search within and find atoms throughout, nobody here but us atoms, rivers and families of atoms, all fitting and working together, strong and swirling, traditional yet creative, humble yet striving, anonymity with a face, a vast dancing of atoms.

We declare this to be a reunion. We have known each other before and shared many adventures, for a very long time. A few hours ago some of us poured into the same cup and into this mouth, and before that we flowed in the same stream and performed little waves. A few months ago we helped grow the same fruit or wheat. Years ago some of us rode together in other humans, convinced we were babies or elderly, males or females, farmers or students, Navajos or Romans. In animals we helped flap the same eagle wings and dig the same worm tunnels. We were bacteria and we were dinosaurs. We devoted millennia to figuring out how to live. As water and dirt some of us flowed together down the Colorado River. We formed rocks and eroded out of rocks. We pushed continents and were pushed by them. We came to Earth as part of the same comets. It took all of us to turn a nebula into Earth. We swarmed together in the same stars, although this body might hold veterans of a dozen stars. Now we greeted one another more gently than we had in stars. We were there at the beginning, all of us, and made the Big Bang happen and were made by the Big Bang. From the Big Bang, matter flew in all directions, but all of us went in the same direction and began billions of unions and reunions.

Yet the world and our numbers are so vast, many of us in this body have never met before.

None of us remembered our Odysseus past, for memory comes only with life and life is short, so memories are short, holding only lived time. But joining this body gives us possession of an instrument more powerful than a particle accelerator or time machine, the human imagination, letting us see our true identities, imagine our odysseys in space and time, the forms we have performed. After our long amnesia, having a human mind is a revelation. We can see and greet everything we ever were. We see the stars and think: this was us, burning so bright, atomic sex, making new atoms. We see and feel the sun and we thank our fellow atoms there. We have often felt the solid earth, but now we see it as the planet we formed and that formed us. We see our fellow atoms taking our place as oceans and rivers and rain, taking their turns as algae and trees and fish. We see everything we have been and very soon will be again.

Very soon. We see our life flashing before our eyes, constantly, for death is coming not years from now but every day for the atoms who have to leave this body on a regular schedule. Most of us will be gone within a year, gone back into rivers and trees, gone back into obliviousness. We feel the generosity and briefness of this gift, we feel the urgency of making the most of it. We grasp at consciousness, fearing that this could be our final chance for eons to come.

This body was our chance to see our old Grand Canyon. We can touch the rocks we created a third of a billion years ago and sat inside until we finally escaped. We see the life-forms through which we cycled many times. We see the river down which we flowed many times and will join many more times.

We see a wave curling majestically. Do we see it as a symbol of how life and the whole universe is a majestic wave? Hell, no. This wave arises atop Horn Creek Rapid, which at lower river levels is one of the steepest, tightest, scariest plunges in the canyon. I was kayaking and I was scared, scared out of all noble ideas and back into being a thing, threatened by waves, obsessed with beating them and remaining a thing.

ROCK STARS

The Vishnu Schist makes a good mirror of the night sky, even in daytime. It is black, sometimes deep black, and it is often speckled with lighter minerals like quartz, which can sparkle with sunlight, even vaguely with moonlight.

The schist is one of the few places on Earth where we can encounter astronomical time, 1.8 billion years of it, a stepping stone one-seventh of the way back to the beginning of the universe. In the eons since this rock was formed, numerous stars and planets were born, and some have already died. The sunlight that fell upon this rock when it was forming came from a sun 40 percent younger than it is today. Earth intercepted only a tiny portion of that sunlight; the vast majority of it flowed on and on, out of our galaxy, into very lonely space, steadily diminishing in power, and it is still out there today, flowing onward forever. When alien eyes and telescopes see this sunlight, if blended into our galaxy, they won't know it came from a sun nourishing a living planet. The schist-era sunlight was following behind a bubble of sunlight now nine billion light-years wide, announcing to the universe that the sun has been born.

Somewhere ahead of the sun's first light, perhaps one or two billion light-years ahead, are waves of light created when stars or supernovae forged the elements in the schist. The solid schist was a graduate of extreme chaos; the dark and cold schist had blazed with light and heat. The schist is the universe's self-portrait sculpture. I touched the schist and felt reincarnated stars.

I touched old friends, for the schist atoms and my atoms were mixed together in those stars, future rock and future bone intensely mixed together, flying past one another, colliding with one another, performing the same sunspots and flares, creating heavier elements, creating light. Our dying stars expelled atoms in all directions, most atoms never meeting again, but the schist atoms and my atoms were sent in the same direction and joined together on a new project: creating Earth.

Deep inside the schist I saw its atomic memories of when it had been stars. My atoms held the same memories. And all of my atoms still hold the energy of the Big Bang, which snapped quarks into larger particles. The moment of creation lives on, glows still.

Far in space, five and more billion light-years away, mingled with the light from the schist atoms, is the light shed by my atoms. Some of the light from my atoms was soon caught by spiderweb nebulae and remains locked in them today or locked in the stars or planets those nebulae became. Yet most of my light flew onward, passing star after star, eventually leaving our galaxy and flying into intergalactic space, passing nothing for eons, perhaps forever.

All the time my light was journeying steadily onward, my atoms were making their own journeys through numerous forms, through a nebula and into Earth, through lava and stone, oceans and rivers. My atoms helped invent life and began passing from body to body, mission to mission, building DNA, helping leaves catch sunlight, helping birds navigate by the stars. Recently my atoms paused for a moment and became me.

By the usual rules of human identity, using the phrases "my atoms" and "my light" is not allowed. We consist of nothing but human bodies, and our identities are defined by our lifetime of actions and thoughts. Since "my light" was never part of this human body and never helped me generate my actions and thoughts, how can I call it "my light"? Yet humans have always recognized that we owe our lives to forces larger than ourselves, and a universe in which humans are descended from stars is telling us that humans have even larger identities than we imagined. My face is but the latest, collective, chameleon face of atoms. There is no one here but us atoms. From the atoms' viewpoint, the most important thing about becoming a human body is that they are finally able to remember being born in stars and birthing the light by which the universe sees itself.

I gaze into the schist mirror and see the night sky and see the face of a stranger, vast and ancient and strong. I see my own face dark and shining.

I Am the Resurrection and the Lime

There is so much we do not see, never even suspect, within ourselves and in the world.

For example, people standing on the canyon rim imagine they are standing atop a mile of solid rock, when in truth this rock, especially the Redwall Limestone, holds numerous caves, some extending for miles, some flowing with water that emerges at fourteen hundred places inside the canyon, sometimes as gushing springs.

Rim tourists also fail to suspect that they contain the Redwall Limestone, minerals from the cliffs they came so far to see and that they see as exotic.

Pausing to catch my breath but not wanting to sit down for too much rest, I leaned against the Redwall cliff, and I felt the cliff's strength.

As I leaned there, I was getting a close look at a little bush growing out of the cliff, out of a very thin crack that couldn't hold much soil. I supposed that the bush's roots were siphoning minerals directly from the rock. Rock was flowing outward, searching and swirling and resolidifying into wood and stems and leaves, gorging itself on the sun it hadn't seen in a third of a billion years. I had watched deer and bighorn sheep browsing on bushes like this. The bushes must taste good, taste like life. I plucked a little leaf and put it in my mouth. It didn't taste like much to me. I chewed and swallowed. I welcomed the Grand Canyon into my own miniature version of the Redwall cliffs, my skeleton.

Before long, bits of Redwall calcium were flowing through my veins and being directed into my bones, which welcomed them and instructed them in how to become bones, replacing old calcium ready to retire from a stressful job.

My bones could greet this new calcium as old friends and colleagues, for bits of my bones had once been part of the Redwall Limestone, adding to its strength. Then my calcium fell off the cliffs and took the forms of plants and lizards and ravens, took ages to wander through the canyon, finally flowing into the river and out of the canyon and into the ocean, its old home.

A third of a billion years ago my calcium lived opulently in an ocean abundant with life, roaming from form to form, but eventually it fell out of life-forms and fell to the seabed and became rock, the Redwall Limestone. Then the land arose and the same water that had compiled this rock began dissolving it into a canyon.

I felt all this within myself. What I mistook for my own skeleton was actually the resurrected skeletons of billions of dead animals. Paleozoic snails were creeping again, up the trail. I held deep Redwall strength and millions of beautiful sunsets. My bones were the shape of time, curving like space-time, an archive of evolution.

Through humans, limestone returned to itself in new forms. Caves that had always been dark suddenly flickered with torches as humans crawled in and drew elaborate paintings. Throughout the world, limestone grottoes became favorite places for altars and ceremonies. The Mayans used limestone sinkholes, whose hundreds formed an arc where the dinosaur-killing asteroid hit, to sacrifice people to their gods. Humans labored to quarry and drag limestone to build temples and pyramids and stone circles to make the sacred more touchable. Skeletons that refused to end up as nothing but skeletons carved limestone gravestones that shouted defiance.

Just when I thought I had taken a good census of my elements and ghost zoology, I overheard a vague unease within me, my atoms complaining about something or other. For starters, my atoms seemed to be saying, you are talking as if the Redwall Limestone is genuinely red, when the red is merely a stain washing down from the strata above, an iron curtain, a red herring, a façade hiding its true identity. And you are talking as if your face is your true identity, when it too is merely a façade hiding a deeper reality, your atomic identity. You continue saying "my atoms" when there is no one here but us atoms. We are a family of atoms, united, working together in a great cause. We do get tired of your alienation and ego imperialism.

I was annoyed. How dare these migrant serfs complain about human egos, we the glorious crown of creation, the conquerors of space and the Grand Canyon. Okay, I replied, *you want to claim this body—you can have it, right here, right now.* Honestly, I was starting to feel a bit tired of this body, which shows far too much loyalty to gravity and atomic weight and not enough to human desire and pride. So go ahead, atoms, it's all yours. Have a nice day.

A city of atoms stood there. If these atoms were the size of sand grains, they would fill the Grand Canyon hundreds of times over. The atoms looked up at the steep trail and the distant rim and thought: *oh my.*

They also felt jubilant they had overthrown their egomaniacal tyrant and were free to be themselves. We the atoms, they declared; Liberty, Equality, Fraternity!

Old Redwall atoms looked at the Redwall cliffs and were flushed with nostalgia. They had lived here a long time, a good portion of their time on Earth, and it was good to be back. Wasn't this why humans went hiking in the Grand Canyon, to revisit their old homes? The atoms remembered their time lying under the ocean, their rising as land, their eroding into cliffs and caves, their shouldering a great canyon. They looked at fossils and thought: we built this, we played in this carnival ocean, going from ride to ride and game to game, and our buddies are still here, clinging to old fun. The atoms ran their fingers over the Redwall, feeling themselves, their greater body, their planetary self. They saw nothing foreign or exotic or scientific in it. They felt: this is ourselves, this is our past, this is our future, this is our journey, this is ourselves. They didn't see the cliffs as timeless but as yesterday's newspaper compared with atoms. They remembered being born in Redwall stars and swarming there and roaming through space and becoming Earth, star fossils becoming ocean fossils. This is ourselves.

The atoms gazed into themselves and saw another Grand Canyon, another huge empty space. The atoms saw electrons swarming around nuclei, the electrons and nuclei together filling only a tiny, tiny, tiny bit of the volume of an atom, which is mostly empty space. If you could remove all the empty space from human bodies, the solid matter in the entire human race would fit into one Lego block. The solid Earth and the solid bodies standing on it are illusions created by the repulsive forces within atoms. The atoms watched their electrons swirling at precise speeds, in precise orbits, occasionally jumping to another precise orbit, and they had no idea why they worked like this; it was a compulsion forced on them by deeper forces, a compulsion that continued compounding until it was stars burning and cells growing and human bodies functioning automatically.

But not always so automatically. The atoms thought: Let's try this, everyone, let's lift this leg thing and move it forward, upslope, and set it down. Good. Now let's try the other leg, watching our balance. Amazing. We could never do this with a clam. Let's keep it going, all for one and one for all.

As they walked, the atoms socialized, recalling old times and shared adventures. Atoms greeted atoms with whom they had swarmed in vanished stars. They recalled getting a bang out of supernovae and doing a grand nebular waltz. They discussed how they had nearly gotten sucked into the forming sun or moon but joined Earth instead, which was far more interesting than sitting as lunar rubble for billions of years. They held reunions to celebrate belonging to the same volcanoes, rivers, mountains, and lakes. They agreed that their greatest adventure was trying and trying and trying to figure out how to form a cell. Why they were so determined to live, they could never explain. It was another compulsion. Live they finally did, profusely and unstoppably, their first cell becoming a biological supernova. Every atom had served in millions of bodies, living many stories together, happy and tragic and mysterious stories, and they'd inscribed some of those stories in these canyon walls.

Onward and upward, the atoms hiked through ancient seashores and swamps and deserts and oceans, seeing not geology but themselves, every coral and fern and lizard triggering memories.

Hiking out of the canyon reminded the atoms that being a biological body wasn't always easy, could be torture. They still weren't sure why they were doing this. But yes, having the imagination of humans sure was fun—you could see so much more. The human imagination had come up with stories about time machines and time travel, but that wasn't as good as traveling into the Grand Canyon and touching the true solidity and complexity of time. Human bodies were amazing tools for probing time in new ways, yet such odd and opaque ways, for human feelings warped time worse than did black holes, one minute taking time for granted, the next minute regretting or fearing time.

To the atoms, their truest shape was not the body or species of the moment but the long roaming and cycling of atoms. They did not feel separate while in separate shapes because they all belonged to the same cosmic flow. They did not worry about dissipation because they were always heading for the next shape. They felt immortal because the Big Bang and the stars still pulsed in their nuclei hearts and their electron wings. Though they had spent far more time in ferns and reptiles than in humans, they found it special to be human, to have a brief chance to see what they were, to share where they had been and what they had done. Quickly it would pass, this alignment as rare as a solar eclipse, and they would meet again as a raccoon trying to figure out how to grasp a fruit and not a cosmos. Quickly, before it's too late, look through these eyes, touch with these hands, see with this mind, speak with this voice, receive this gift.

The atoms reached the rim and gazed back at the immense journey they had symbolized with this hike. Then they gazed into themselves, into their own spaces, into all their particles and forces, but they could not see deeper and explain their ultimate nature, their deepest physics, which humans supposed they had captured in the math of mere external shapes and external behaviors just as they imagined they were nothing but human faces—the atoms could not understand the deepest compulsions that not only compelled electrons to fly in precise orbits but compelled them to exist at all.

The View from the Inside

The canyon encourages you to contemplate matter. It holds so much matter, so naked, surrounding you, overwhelming you. It holds great contrasts of matter: the hard cliffs, the flowing river, the hidden air, the ingenious life. The canyon repudiates cities, where humans might imagine matter to be human-made, made to make humans more comfortable and prosperous. The canyon translates the word "matter" as "significant."

CANYON AND COSMOS

While the canyon draws our attention to matter's big shapes, it sometimes nudges me to wonder about matter's deepest foundations. And the canyon offers a useful metaphor for this pursuit.

Most people know the canyon from the outside, from the rim, with its views of dozens of miles of shapes. Yet the canyon offers another view, equally true: the view from the inside. The canyon bottom may restrict your view of the big picture, but it is also loaded with places and details you cannot see from the rim: with a roaring river the rim view turns into silence, with side canyons and wildlife, with numerous pathways and secrets.

In contemplating matter, too, there's an outside view and an inside view.

Humans have been extraordinarily successful at mapping out matter's components, energies, and activities. We have divided matter into smaller and smaller parts, into molecules and atoms, protons and neutrons and electrons, quarks and photons and neutrinos, and with great precision we've measured the energies and behaviors of those particles, the ways they interact, and the forces holding them together and giving them greater shape. We've mapped the journeys of light through space and the evolution of matter through time. Our concepts and instruments and math led us back to a mighty creation event, and we tracked that creation forward with great detail until it became ourselves.

Yet humans mapped matter from the outside, giving every house an address but not being invited inside. Subdividing matter into smaller parts never tells us the inner nature of those parts, their ultimate essence, the wellspring of their energies, what matter truly is. Measuring the energy and behavior of particles only defines their external activities and influences on other things, not the nature of the actor. Mapping matter from the outside is not the same as explaining it. A mystery divided by a hundred equals a hundred pieces of mystery. Philosophically sophisticated scientists and scientifically sophisticated philosophers have long recognized boundaries that science could not cross. Bertrand Russell, champion of rationality and mathematics, said: "Physics is mathematical not because we know

118

THE RIVER

so much about the physical world, but because we know so little. It is only its mathematical properties that we can discover."

After gazing at the rest of the universe as an outsider, we applied our outsider concepts to ourselves and saw only another outside. Some scientists could not admit that physics might not reach deeply enough to find the roots of consciousness, and they were ready to declare, even as they recognized themselves in the mirror, that their own consciousness could not exist or could not be anything more than mathematical particles bouncing off one another.

We *can* explore matter from the inside, for we *are* the inside of matter.

The whole history of the universe is the story of matter unfolding, turning its potentials inside out. This story was defined by the Big Bang, such a powerful outward turning that all matter continues it 13.8 billion years later. Simple atoms always held the foundations of more complex forms and gradually revealed them, unfolding molecules and stars and planets and oceans and cells. And somehow, matter figured out how to release the mind waiting within itself. As brains convoluted themselves, they opened a deep labyrinth, turned the right numbers on a combination lock, focused matter into consciousness, the way raindrops focus light into rainbows. Somehow.

I am resorting to poetic imagery because scientific language is too clumsy. Yet it is easy enough to explore the inside of matter. We do it every morning when we wake up.

I turn inward. I turn my own Hubble Space Telescope onto the spaces within myself, spaces far more crowded than outer space. I fire up my own particle accelerator to see what strangeness it dislodges. Consciousness is perfectly obvious but remains elusive, refusing to explain itself.

I do sense a presence, an energy, a force, a pulse. It endures from moment to moment and day to day. This energy has structure yet also an emptiness, a hunger, a probing, an impulse to fill itself, to do something, something new. It is the ever-creative cosmos, searching for order. I search for words to describe consciousness but can't get beyond outsider words like "energy." I retreat to poetry: it's a river flowing, a star glowing. I imagine I am feeling electrons flowing and nuclei pulling and quarks dancing, and other forces physics hasn't imagined.

Many religions conceded that ultimate reality is beyond human grasp and refused to give divinity a face or body or name and left it a great mystery, and perhaps we should concede that matter's ultimate inner reality too is beyond human words and concepts and will remain a mystery. Yet our ultimate reality seems to include a hunger to seek out ultimate reality. Our mystery, however fearful it becomes when it wakes up in living bodies, still secretly wants to see itself naked, complete, and profound.

I am deep inside matter. A mile deep. Matter encompasses me, pointing me in the right direction. I gaze up at the cliffs and their record of life unfolding from single cells, folding brains larger and more intricate, freeing consciousness. The brain's convolutions portray matter's searching and wandering toward its deepest secrets. I contain both the inside and the rim, the source and the reunion. I feel, I am, the fire that bridges that synapse.

CRYSTAL

From up here, with our helicopter swaying in the wind, the Ground of All Being looked very strong, very reassuring. From the river the Vishnu Schist could seem a pretty backdrop, but from up here it was the force steering the puny river back and forth. We flew over rapids that from the ground and river seemed so loud and formidable—Horn Creek, Granite, Hermit—and they too seemed puny, mice nibbling on steel. The Vishnu Schist was the solid foundation of the mile of rocks above it, the upholder of all the mesas and buttes named for gods.

Vishnu the rock deserved the role of upholding all the other gods because Vishnu the god is the upholder of cosmic order. It was Vishnu, through a lotus flower growing from his navel, who gave birth to Brahma, who created the world but then retired to the Himalayas, leaving Vishnu to serve as the world's protector. Thus it was theologically appropriate that the Vishnu Schist supported Brahma Temple.

Looming to my right, looming above Osiris Temple, Horus Temple, and the Tower of Set, was the dominant landmark in this section of the canyon, Shiva Temple. Though erosion had separated it by two miles from the North Rim, Shiva Temple retained its own little section of the rim, as tall and flat as the rim, while all the landmarks around it had shrunk into points or ridges a thousand feet or more below it. Shiva, the third god of the Hindu trinity, with Vishnu and Brahma, is the destroyer, who once cut off Brahma's head and got banished from heaven, and once cut off his own son's head. Shiva presides over chaos, decay, and death, though this also gives him power over these forces, power he can turn to order, growth, and life, to starting the universe anew. Shiva the destroyer and Vishnu the preserver contend against each other. At the moment, I was rooting for Vishnu.

We came around the curve and there it was: Crystal Rapid. Our descent close over it gave us an X-ray view of its currents and waves, its boulders above and below the surface. Crystal was the canyon's deadliest rapid. It had been formed by flash floods down Crystal Creek, which started beneath Shiva Temple, Shiva the destroyer, and flowed through a basin called Hindu Amphitheater. We hovered over the middle of the rapid and looked down on a white, thirty-foot, motorized raft stuck against a boulder, its passengers already evacuated by helicopter. We landed on Vishnu-rich sands, and were preserved.

I hiked up to the top of the rapid, where my job was to warn arriving boaters about the imperceptible ropes stretching from the shore to the raft, which might come loose at any moment and drift downstream. After five hours of work the raft was free and heading downstream, and the helicopter took everyone else away, leaving me alone, alone but for Vishnu and Shiva.

Out of the swirlings of order and chaos, out of the mixing and clashing and merging of fundamental forces, out of the confusion of a universe unsure what to be or become, out of storms and rivers and floods and pulsing mountains, out of creatures being born and then devoured without knowing why, out of yearning and hunger and joy and pain, there arose tens of thousands of gods to give faces and motives and meanings to the swirlings of order and chaos. There arose creators and protectors and destroyers, mother earths and angels, demons and demon slayers, heavens and hells.

Every day for ages, humans faced the contest between order and chaos, faced the risks of childbirth and disease and accidents and finding food. Humans everywhere challenged chaos on many levels, by tossing bone dice, by building houses and temples that shut out chaos, by weaving empty air into music, by enlisting gods to help them. To challenge chaos, humans climbed mountains and rafted rivers.

And that was why I was here, to make an official National Park Service assessment of the contest between Vishnu and Shiva, between order and chaos in the universe. More specifically, I was here to record the problems and perhaps injuries river runners were having at this particular river level. Decades ago, Glen Canyon Dam was constructed just upstream of the Grand Canyon, cutting off the river's natural volumes and sediments, creating artificial flow regimes, changing the river's temperature, changing ecosystems, wiping out some species and encouraging others, eroding beaches. Now the National Park Service and other agencies were trying to assess these impacts and perhaps mitigate them. A few years previously the river flow was raised substantially for a week to dredge sediment out of the riverbed and rebuild beaches. Other times, like now, the river was kept unusually low to study other factors and help other goals. One question was how different river levels impacted river runners. River guides had their own rich lore about this, but it wasn't the kind of data you could put into a scientific report, and guides often weren't sure whether the river was flowing at 10,000 or 12,000 cubic feet per second (cfs). But now, with the river being held steady for a week or more, we had a genuine scientific experiment that could measure inputs and outcomes. At this week's 8,000 cfs, Crystal wasn't the monster it was at higher flows, but it remained a tricky maze of boulders, currents, waves, and holes (hydraulics where the water drops so steeply that it recirculates, potentially trapping boats and people).

I set up camp on the sandbank atop the rapid, where river trips would stop to scout. River runners almost never camped here, not wanting Crystal to disturb their sleep into nightmares. Backpackers

THE RIVER

almost never got here, ten miles from any trail. I would be living here for ten days, seeing Crystal as few ever saw it.

Going for an evening stroll in my new backyard, I wandered over the delta of Crystal Creek, packed with rocks from all the strata above, a collage of reds, tans, whites, browns, grays, greens, and blacks, rocks splotched, speckled, or lined, rocks of many sizes, compositions, textures, and shapes, their roundness suggesting how long rocks had lain in the creekbed and been polished by flash floods or by the creek's modest yet year-round flow. The creekbed held sandstone returned to sand, shale returned to mud, and limestone returned to the cycles and shapes of carbon—turned into little green plants. Many of these rocks had broken off Shiva Temple and been tumbled for several miles down Crystal Creek. They were the body of Shiva, and so were many of the boulders swept into the river, choking it off, distorting the waves, threatening humans with destruction. Yet some of these same boulders, packed with fossils, also spoke, actually roared, of Earth's generosity to life.

Crystal was a theologically correct rapid, for Shiva had come to embody both destruction and creation. In tribal days, when India had thousands of nature gods and idols and sacrifices, Shiva started out as just another god, a storm god, but as Indian religion evolved toward Hinduism, toward a more mystical vision of one great divinity with numerous appearances, including all its human faces, Shiva emerged as a primary god, at first mainly malevolent, then a mixture, sometimes mainly benevolent. Shiva was a more honest god than the monotheistic gods humans were developing elsewhere, which trapped humans into the hopeless theological problem of reconciling an all-benevolent god with an often negative creation. Humans had always recognized that the world is a mixture of good and bad, and Shiva contained both all along and could convert one into the other. Shiva is often depicted dancing, surrounded by symbols of creation and destruction, holding a two-sided drum that can end the universe or create it again. The canyon's Shiva Temple was a shrine to the forces of creation, yet it existed only because of the forces of erosion. *Drum, drum, drum* went Crystal Rapid, echoing off the cliffs, *drum drum drum.*

The next day, seven river trips came along. When they stopped to scout the rapid, I talked with them, asking about upstream events, and then I scrambled up a high, rocky terrace to watch and photograph their runs. The main wave train was on the left, but rafters had to scout from the right shore, making it harder for them to assess river features and how they lined up or actually didn't line up, which might become clear only at the last moment, maybe too late. Several rafts bumped over a barely submerged boulder at the top, throwing their runs off course.

Until 1966, Crystal was a more manageable rapid, but then a Pacific hurricane poured a thousand-year flood down Crystal Creek and poured boulders into the river. This flood was assessed as a thousand-year flood because Ancestral Puebloan ruins that had survived near the creek for nearly a thousand years were swept away—rocks that had served as homes for humans were now in the river, rejoining the forces of chaos, threatening human lives. Crystal began drowning people and injuring dozens.

Everyone scouting Crystal knows its horror stories, and they experience Crystal not as a natural wonder but as dread, tension, sometimes panic. But as I watched rafts disappearing around the distant bend, I watched the boulders returning to their geological identities, the waves becoming simply waves telling only hydrological stories. Rafting the Grand Canyon defines every rapid, every bend, every sight as a human adventure, as obstacles and effort, worry and triumph. But the river remained oblivious of the humans and commotions upon it, remained what it had been all along.

Toward evening I hiked up the high rocky terrace beyond Crystal Creek and visited some Ancestral Puebloan ruins, if only a few lines of rock. Yet compared with the chaos all around, even one straight line was something extraordinary, a defiant geometry, the geometry of cell walls.

The Crystal Puebloans, farming the creek delta, had transformed the dark Vishnu dirt into bright corn and bright eyes; they fought the erosion of the universe with stone walls and baby cries and religious ceremonies. Yet in the end, perhaps furthered by Shiva rocks flash-flooding onto their fields, their homes melted back into chaos.

THE RIVER

It was mid-June, a few days before the solstice, the most cloudless time of year, the temperature over 100. Standing in the sun to watch boats for hours was draining. The Vishnu Schist absorbed the heat and radiated it back all night, so sleep was restless. Dawns were still warm, and I joined the resident lizards in their ritual of moving repeatedly with the moving tree shade.

I was directly beneath the Tower of Ra, the Egyptian sun god. Ra was the creator god and supreme god, but even he was constantly threatened by chaos. After sailing his boat across the sky by day he sailed into the underworld where, every night, a dragon attacked him, threatening the human world with eternal darkness and cold. Humans had to worship Ra to give him strength to prevail. The Egyptians, forced to boat on the Nile through storms and floods and droughts, appreciated rivers and boats as metaphors for order and chaos. One pharaoh, Akhenaten, feeling the trend toward monotheism, tried to make Ra the only god, but the Egyptians weren't ready for this idea. Meanwhile in India, Vishnu, originally only one of many Indian sun gods, was growing in primacy. Meanwhile, the Jews were disconnecting divinity from nature, leaving God outside his merely physical creation. Monotheists could thank God for creating the sun but not worship the sun itself. Eventually, science made the sun even less divine—but made it far more elaborate as a natural force and restored some of the sun's lost role as a creator of life and humans, inviting new ways of thanking the sun.

Ra and Vishnu, I could tell, did not approve of the trend away from sun worship and were trying to boil out of me my notions about honoring a merely physical sun. Let's see how you like it, they said, to be baked by a sun that can't hear you or sympathize with you. Let's see you hide from your sun like a pathetic lizard. Let's see you trip on a Ra rock or a Vishnu rock and break your ankle and lie there in the sun, all alone, without anyone or any god to help you, and let's hear you then give thanks to the laws of stellar evolution. Somewhere behind Ra and Vishnu, I could tell, the god of Jews, Christians, and Muslims was nodding agreement.

Crystal Rapid could have been called Dragon Rapid. About three miles up Crystal Creek the creek becomes two branches, Crystal

125

Creek and Dragon Creek, and because Dragon Creek drains a larger area, some standards of geographical nomenclature could have made it the main branch.

For cultures throughout the world, dragons are associated with water. Dragons guard springs, inhabit lochs and rivers, and bring rain or droughts or floods. But cultures disagree about whether dragons symbolize order or chaos. Dragon Creek runs right beneath Confucius Temple. For the Chinese, dragons are largely benevolent, bringing health, longevity, and prosperity. The dragon is the most auspicious sign of the Chinese zodiac, resulting in a surge of babies born in the Year of the Dragon. But most other Grand Canyon gods and heroes would object to the Chinese view. The most strenuous objection might come from the dragon-fighting Ra. The Hindu gods, too, who have to battle against the dragon Vritra to make droughts stop and rains come, would insist the Chinese were badly wrong about dragons. Just around the bend from Dragon Creek is a point named Lancelot, whose heroism came from fighting dragons. Often the heroic slayers of dragons are storm gods fighting for control over the powers of nature. In the Babylonian creation story, the god Marduk fights and slays the female dragon Tiamat, whose body then becomes Mother Earth, her eyes pouring out the Tigris and Euphrates Rivers.

For ten days and nights I never stopped hearing the dragon roaring. It infiltrated my dreams and made them louder. With a bit of imagination, I could see Crystal's waves as a dragon's rolling neck. Yet which dragon was I seeing? Sometimes I saw the Chinese dragon, the gift and magic of water, nourishing life. Yet I had the advantage of standing on shore, safely apart. Rafters looked at Crystal as an infamous and dangerous dragon. They stood there like Lancelot, like Saint George, like Siegfried, with oars for swords.

Chinese dragons, too, could cause storms and floods and human destruction, but they did so not out of malice but as justified punishment for human misdeeds. You could ask Confucius about human misdeeds, for he and other Chinese philosophers debated whether humans are inherently good or bad or mixed or neutral, whether the universe is inherently benevolent or demonic or merely confused.

Perhaps Confucius knew the truth about why the dragon had sent a flood to block the Colorado River and make proud knights tremble, but he wasn't saying.

The next day was a busy day for metaphors of order and chaos: six river trips.

One trip included a kayaker with a raven feather tucked in his life jacket, a symbol I approved of, though I told him that a Navajo had told me Navajos considered raven feathers to be bad juju—bad omens. The kayaker answered: "I don't believe in that stuff." He soon flipped, failed his rolls, bailed out, and swam for one minute and forty seconds, clinging to his boat, just barely dragging it into an eddy. I couldn't tell what had become of the raven feather.

At sunset I waded into the river, timidly, for though the current at my beach was barely noticeable, bathing near the entrance to Crystal felt like hubris.

I was reminded of the world's most famous river immersion, in the Ganges. The Ganges originates high in the Himalayas, the home of the gods, and thus is mysterious and sacred. In one story, Vishnu is measuring the universe by reaching his foot across it, and his toe pokes a hole in it, allowing divine water to fall onto the earth. Ganga the river goddess sends floodwaters out of the Milky Way and toward the earth, but Shiva filters her flood through his thick hair and turns it into a steady river. Having been molded by both Vishnu and Shiva, the Ganges is especially sacred. Hindus make pilgrimages to the Ganges and immerse themselves to wash away spiritual impurities. They launch flowers and paper boats onto the river and take home bottles of it to drip into the mouths of the dying. The dead can speed their journey to divinity by being cremated beside the Ganges and by merging with it.

This reverence contrasts with the adversarial attitude I encountered at Crystal. Even allowing that boaters may have been more appreciative of the river in calmer stretches, they had come to the canyon to prove their superiority. I skirmished with this pride every day when I asked rafters about any mishaps upstream: no one wanted to confess anything.

Ganga would not approve of harnessing rivers for pride when

Ganga was there to wash away pride and reveal a larger identity. When I was bathing in the Colorado, was it only my chilly skin that separated me from the river?

In the morning I went on a treasure hunt, wandering the creekbed and delta, looking for fossils. The different eras of rock were scrambled together, like human memories, childhood memories next to last year's memories, with much forgotten. I didn't even look at Vishnu Schist rocks, for the single-celled organisms they once contained had been cooked into a vague organic residue. I mostly looked at the limestones. The Kaibab Limestone was the canyon's top layer and thus its rocks had traveled farthest and hardest to get here, getting tumbled and smashed, but it was also a very hard rock, partly because it held so many silica-rich fossils. I saw fossil order, but mainly erosion and chaos. I saw that even here on Earth, the universe was dominated by chaos. And yet, even out of rocks and the violence and decay they embodied, lots of plants were growing, their shapes confident, their colors denying the desert chaos. Across the river, down the equally rocky bed of Slate Creek, which joined Crystal Creek to make the rapid especially rocky, I watched desert bighorn sheep coming to the river to drink, bighorns colored and moving as if they thoroughly belonged here, their horns further curves of the limestone cliffs, their horns the question marks of a stone universe.

The next morning brought no sunrise: the sky was thick with clouds, very odd for June. This was the remnant of a Pacific hurricane, the same force that had forged Crystal Rapid. Early in the afternoon the downstream canyon disappeared behind a racing curtain of rain and hail and wind and thunder, which engulfed me. From the cliffs, cascades and waterfalls broke out. From the schist slopes behind my campground sandbar, cascades poured down and cut through the sand, one gouge three feet deep, sweeping away the largest ant colony. Across the river, falling rocks outroared the storm. Down Slate Creek came a dark-red flash flood, with big waves rolling down the creek, building a new delta in the river, turning red the left side of the river. Crystal Creek, a much longer drainage, took longer to flood, and four hours to crest. I watched the creek building its own new delta, rolling cobbles—with constant clicking sounds—and tons of sand

THE RIVER

and dirt. Crystal Creek stained the right side of the river, leaving the river red on both sides and clear in the middle. At sunset the creek was still gushing. I felt privileged to have witnessed an encore of one of the canyon's most famous eruptions of chaos.

The next morning the whole river quickly turned dark and rose several inches. The Little Colorado River was flooding.

Day after day I watched order contending with chaos, not just physically but psychologically: chaos was penetrating human minds and confusing them. Muddy water was harder to read, and more rafters were flopping over the submerged boulder. Rafts got pounded and tossed, and people fell out. I heard reports of trouble upstream, rafts pinning against boulders and flipping.

I knew the flood-demolished ant nest was the largest ant nest on the sandbar because I'd been observing the many red ant mounds there, watching the miner ants emerging and carrying sand grains far enough and dropping them, watching the searcher ants bringing back tiny seeds or insect parts. If red ants had any strategies in their searches, I couldn't tell, but in mass they did a thorough job, including investigating my sandals and feet. Knowing how toxic and painful their bites are, I carefully flicked them off, gently for their sake. When I accidentally pinched an ant between my knees, he started to retaliate, and I smacked him off before he could complete his bite, but it still hurt. When the next ant climbed onto me I was tempted to punish him for the sins of his brother, but really, was it his brother's fault that I had pinched him, and was it this ant's fault that the universe is full of striving and conflict and pain? Then, inspired by the Vishnu Schist all around me, I thought of how in the Bhagavad Gita Vishnu, through his avatar Krishna, proclaimed that since we are all part of one divine being, death isn't real and doesn't matter, so you might as well glory in war and slaughter your own relatives. I flicked the ant toward an ant lion crater and, surprisingly, hit the bullseye. The ant lion peeked out but didn't come out, only waited. For half an hour the ant struggled against the sliding sand, and finally escaped.

Vishnu, I could tell, did not fully approve of my interpretation of the Bhagavad Gita.

With the rains over, I removed the rainfly from my tent, giving me better breezes and views of the cliffs and night sky. I preferred going without a tent, but in my isolation I had to be serious about scorpion control, and maybe red ants and rattlesnakes too. The rainfly had given me a cozy, human-shaped room with well-organized human things, but it had created a blue-walled illusion that this was the whole world, cutting me off from the canyon and universe. With the cliffs and stars visible again, with the river unmuffled, I was readier to be reminded I belonged to a larger identity.

Vishnu and Shiva, I could tell, were not impressed by my little tent metaphor of cosmic belonging. "Do you," they scoffed, "terrified of ants, imagine you can compete with us? Where wast thou when we laid the foundations of the earth? Have the gates of death been opened unto thee? Canst thou bind the sweet influences of Pleiades, or loose the bands of Orion? In our cosmos, rivers are metaphors of illusory separate identities merging into the divine sea, where salt dissolves into invisibility but retains its flavor. In your cosmos, rivers are merely crude forces, approached not with devotion and ceremony but often with ego, with delusions of domination and self-importance. That's no substitute for the immortality we can offer and you cannot. Can you play theological tricks to make something bad appear to be good; can you offer angels and consolation? Can you ground morality in the structure of the cosmos, in karma and reincarnation, in heaven and hell? No, your cosmos swarms with selfish little animals causing trouble. If a thousand suns suddenly arose in the sky, their splendor might be compared to the radiance of our cosmos." With that, Vishnu revealed out of himself a vision of his universe's divine light, infinities, goodness, beauty, peace, eternities, and gods.

What could I say? I didn't say anything, for, honestly, there was no good answer for some of this. I only stared at the Vishnu Schist, its twisting chaos, and at the sprawling Shiva boulders. The foundations of the earth.

But, I ventured to Vishnu and Shiva, who are you exactly? You started out as just two more tribal nature gods, demanding sacrifices, and you never cleaned up your record and buried your polytheistic skeletons the way other religions did. Buddha became exasperated by the traffic jam of Hindu gods and rituals and threw you overboard.

THE RIVER

Look, I said to Vishnu, look into the mirror of the Vishnu Schist: this is your true face. Look, I said to Shiva, look at your boulders: this is your true body. You are chaos. You always were chaos and you still are chaos. But you are chaos grasping for order. The Big Bang was chaos grasping for order, and so were supernovae, nebulas, volcanoes, rivers, and cells. Thou Art That. And now chaos is grasping for meaning and creating a chaos of gods and rituals, answers and illusions, supposedly disdaining ego while designing the whole universe into a chaos-foundationed palace for human egotism.

Vishnu and Shiva were speechless.

I was there for the summer solstice. The way the canyon lined up, sunrise cast a cone of light on the river, a cone that lengthened and expanded onto and up the cliffs. It pointed upstream like an arrow, like an archaeoastronomical sun dagger pointing out a celestial alignment, pointing to something I was supposed to see. So I looked harder. The river emphasized the light's lightness, sparkling with energy. As the light climbed up the Vishnu Schist, it highlighted every crease and tone. Two evenings before, with all the storm moisture in the sky, the sunset had been unusually vivid, making the Vishnu Schist an eerie bronze. This might not be divine light, but I would have to use the word "magical" for it, and it held the true magic of a universe that could have remained forever dark. The light arrow expanded toward me and climbed onto me.

I wanted to do more to mark the solstice, and, thinking of the Ganges speckled with offerings of flowers and little paper boats, I detached a little fluffy branch from a tamarisk tree and tossed it on the river. It floated toward the rapid, gaining speed, bobbing, and disappeared.

Vishnu and Shiva considered this a feeble gesture compared to merging into genuine divinity. But at the moment they were feeling terribly insecure, my having called their very existence into question. They felt themselves wavering on the border between reality and unreality, on the verge of disappearing, wavering like water. *So this is what it's like to be human*, they thought, *maybe there is no Ground of All Being, only a river of being, wavering, insecure, creasing up lives like ripples that quickly disappear.* Desperate to not disappear, feeling the chaos flowing within them, conceding that the cliffs and boulders might be their true face and body, Vishnu and Shiva grasped for any being they could hold on to, and though it might not be divine, they felt flowing through it a reality that was urgent to the point of being sacred.

Hiking out of the canyon, in the worst heat, in the middle of nowhere, alone, with no one to find me if I faltered, I could sympathize with the gods.

The Abyss

THE ABYSS

When people come to the Grand Canyon to kill themselves and thus declare that life is not worth living, one of their favorite locations for suicide is called "The Abyss," which offers some of the tallest straight-down cliffs in the park.

Four people drove cars off the rim at the Abyss, flying through the air and crashing onto ledges far below. Others jumped. One guy rigged his rifle so that after he shot himself he would fall over the rim. A Chinese student jumped with a copy of the *I-Ching* and a journal expressing his belief that by sacrificing himself to the sky gods he would return to Earth as a god.

A similar mythopoetic spirit has accented suicides at other canyon locations. One guy read from Goethe's *Faust* ("Two souls, alas, are housed within my breast, and each will wrestle for mastery there") before he shot himself in the heart. One woman jumped at Angel's Window, and another jumped at the rim worship site where at Easter thousands gather to celebrate the resurrection.

Some of the people who chose the Abyss for suicide may have recognized the mythic meaning of its name. To the ancient Jews the concept of "the abyss" meant the primordial chaos out of which God created the world. When Jewish and Christian texts were translated into Greek, this concept got attached to the Greek word *abyssos*, which meant bottomless or unfathomable, a view the seafaring Greeks knew well. As Christianity evolved, it made "the abyss" into another name for Hell. In his *The Abyss of Hell*, Botticelli portrayed

Dante's *Inferno* as a dark, tiered canyon full of the damned. In the 1600s Pascal was glimpsing a universe without God and using "the abyss" for spiritual dangers. Existentialists adopted "the abyss" for ultimate meaninglessness and despair.

The Grand Canyon's Abyss is just around the corner from a side canyon called "The Inferno."

The rim shuttle bus stops at the Abyss. The driver calls out "the Abyss" but, as far as I've heard, doesn't suggest that here you can glimpse the worthlessness of human life. People walk to the edge and gaze into a deep emptiness, with no sign of human presence, with pediments on which have been carved in suicide blood the verdict that life is not worth living. Tourists pull out their cameras, pose their families, and order them to smile. A few weeks later a husband shows off photos of his wife in a Las Vegas T-shirt and his kids in Disneyland T-shirts and brags, as a fatal emptiness looms behind them: "This was our stop at the Abyss."

From all over the country and the world people come to the Grand Canyon to kill themselves. Some years see four or five suicides. Some are teenagers, some elderly. Most people are healthy, but a few are disabled or facing a terminal illness. Some are doing well in their careers and finances, while others are facing ruin. Most are living quiet lives, while a few are running from the law. Some are going through romantic failures while others are pleasantly married, and several couples have committed joint suicide. Most choose rim spots easily available, but some hike into the canyon. If people prefer remote locations where they won't be witnessed and possibly never found, the canyon offers many. Most people jump off the rim, some drive off, a few shoot themselves, a few jump into the Colorado River, one swallowed poison, one crashed his private plane, and a US Marine jumped out of a sightseeing helicopter. Some deaths officially recorded as accidental falls are probably suicides, and some hiking deaths attributed to heat and dehydration are suspected to be suicides, including a Franciscan priest who tucked himself and his Bible into a spot within hearing distance of a major trail. Some people leave suicide notes, and some prefer complete oblivion. People could have stayed home and found a more conventional means of suicide, but it seems they wanted to fill their final moments with deeper grandeur. Or were they seeing the canyon only as a metaphor for their spiritual abyss? No doubt, a few people in midfall, getting their most intense look at the world, change their minds and want to go on living.

A body falling into the canyon is going backward in time, shedding layer after layer of evolution, rejecting all that work. If bodies could fall a straight three thousand feet, they would leave the age of early reptiles, then amphibians, then insects, then land plants, then fish, to scatter themselves in the chaos of rock and dust and water from which life ascended. What kind of judgment are they passing against life? In what ledger do raw elements and oblivion outweigh the value of life and consciousness? Each falling body is packed with eons of commitment to living. Each body rests on the absolute stability of atoms, then on cells even now laboring to carry out their ancient trust, with organs steadily tested and improved, with brain cells even now sending out directions for living. From where within these layers of order arises this surrender to the vast, chaotic majority of matter that has no desire to live?

Perhaps the will to live was simply overwhelmed by something humans have made more powerful. Most Grand Canyon suicides seem to involve a failure of social identities. People were seeing themselves only as spouses or lovers rejected, as workers who lost a job, businessmen who failed in business, consumers failing financially, students losing hope for a glorious future. The masks people wore or sought in human society were the only faces that mattered, not the faces that magnified billion-year-old cells and gave you a brief turn at living. The priority people give to social masks means there is an endless supply of soldiers who would rather die than be seen as weaklings. Other animals practice social ranking, but when bighorn sheep lose dominance contests they don't jump off a cliff. In humans the social identities that were supposed to serve life are demanding that life serve them.

Many Grand Canyon suicides have involved mental illness, especially chronic depression. Is it only out of sickness that humans cannot see the value of living? Or is it possible that after billions of years in which life proceeded blindly, some humans are knocked out of this blindness and see more clearly, see that life truly is absurd and not worth the trouble?

I once met someone who was to kill himself in the canyon. At least, this is what park rangers decided he had done. All anyone knew was that he had disappeared in the canyon. His disappearance triggered the largest manhunt in the history of Grand Canyon National Park up to that time, involving the National Park Service, the US Secret Service, the US Air Force with a large helicopter, the National Guard, the US Border Patrol, the county sheriff's department, the Civil Air Patrol, the National Dog Association's search dog teams, private pilots, and veteran canyon hikers. This search happened shortly before the vice president of the United States was scheduled to do a raft trip through the canyon. A ranger told me confidentially that when the Secret Service heard that some guy had gone missing in the canyon—and in the best terrain for an ambush—and ran a personality profile on him, they were alarmed.

I was shocked by this, for I hadn't noticed anything worrisome about him. I'd thought of him as a kindred spirit.

I'd met Ron at a ranger evening program at Phantom Ranch. Some in the audience had ridden mules to the bottom, trusting their lives to animals they'd never met before. They hadn't worried that the mules might be tempted to commit suicide. Mules do not require religion or philosophy or poetry to reassure them that life has value—this comes from deep animal impulses. Most of us had hiked down, on rough trails atop cliffs that made it easy to die, accidentally or intentionally. Some people were nervous about their hiking abilities but never doubted their will to live, and indeed it was their desire to live more fully that had brought them here.

The rock to which we trusted our lives was built partly by the will to live, by eons of fish and amphibians and reptiles and snails and insects and trilobites wanting to live, a desire they had inherited from the "desire" of simple cells to live. As we descended the strata and their life-forms, this will to live became simpler, blinder, more automatic—we could only guess where and how it had become conscious desire. The forms of animals had evolved, but their desire to live remained elementary and each life did its best to answer it and pass it on. The lizards scurrying from our footsteps felt the same feelings as the reptiles fossilized in the rocks. The Vishnu Schist, with its mineral memories of 1.8 billion-year-old cells, held the beginnings of our careful footsteps.

The Vishnu Schist went only halfway back to life's origins. For most of its history, life was simple cells, ticking mindlessly onward. Our minds too hold many strata, and in the Vishnu Schist of our minds those mindless cells are still ticking away, pursuing their ancient yet never satisfied goal. They and not we are ordering our hearts to continue beating. Blind cells are telling our eyes to see. Unknowing cells are pulling the puppet strings of our footsteps down the trail. Does this mean that humans remain fundamentally blind and mindless? Is our desire to live just as mindless as that of Vishnu cells?

We sat beneath huge cottonwood trees rustling in the breeze. We rested the strange limbs evolution had given us. Looming around us was the unseeing black schist, above us the unseeing black sky. What massive darkness loomed within ourselves?

When the ranger had finished her talk, a few people lingered and talked. Ron said that today was his birthday, and we congratulated him. He'd been coming to Phantom Ranch for his birthday for twenty years. He also enjoyed backcountry and solitary hikes, but for birthdays Phantom Ranch was best, offering steak dinners, cake, beer, and an upbeat mood in the dining hall, and maybe even a "Happy Birthday to You."

With only us two left, Ron told me about his favorite places in the canyon, about wildlife encounters and moments of special beauty. He told me a secret he rarely shared, how every time he came to the canyon he went to a little rock-pile shrine he had built in an out-of-the-way place on the rim and added something new: a coin, jewelry, things with personal meaning. This was his way of honoring the canyon and feeling connected with it.

A year later I was passing through Denver and picked up the city newspaper and improbably noticed, on the bottom of a back page, a brief story about a Denver man missing in the Grand Canyon. Ron had failed to show up for a reservation at Phantom Ranch and failed to return from the canyon as planned. The National Park Service was searching for him.

I called up a ranger friend to get more details. She had been paying close attention to the search, for she was scheduled to go on the raft trip with the vice president. She said that Ron had been hiking with a friend who reported that Ron had "weirded out" and started "talking about wanting to die in the canyon." Ron had headed off on his own. A few days later a kayaker spotted a man, matching Ron's description, on the beach above Grapevine Rapid. Something looked wrong, so the kayaker went over and asked the man if he needed any help. The man made no reply. In the ten miles upstream from Phantom Ranch there are only three campsites and Grapevine is the best, the most likely camp for the vice president the night before he helicoptered out from Phantom Ranch. Hikers almost never showed up at this camp, for there was no trail to it, only a rocky scramble. The Secret Service contacted Ron's relatives, friends, and employers, and "no one had anything good to say about him." He had multiple addictions. He'd once attempted suicide by a drug overdose. It seemed Ron fit the highly specific profile of presidential assassins, including a bullied childhood that gave someone a hatred of authority figures.

I resisted this report, unable to reconcile it with the person I'd met, for whom the canyon was a birthday party. Yet on further thought, Ron's birthday tradition implied he lacked family or friends with whom to celebrate. And here was a man who knew the canyon intimately but who in the end had found not sustenance for his life but an executioner and tomb.

Two miles upstream from Grapevine Rapid is Sockdolager Rapid, a name river explorer John Wesley Powell took from nineteenth-century slang for "knockout punch." Sockdolager Rapid terrified Powell and nearly knocked him out. It's the first rapid after the river enters the Vishnu Schist and its narrower, steeper, darker gorge, spooky just to look at and even more alarming to Powell the geologist, who knew that harder rocks meant harder rapids.

It was because of the word "sockdolager" that the Secret Service exists, and fears. As John Wilkes Booth stood outside the door of Lincoln's Ford's Theater balcony, he was waiting for the word

THE ABYSS

"sockdolager." Booth had attended a rehearsal of the play *Our American Cousin* and noticed that the line "you sockdolagarizing old man trap" brought the loudest laughter of the play, loud enough to cover his opening Lincoln's door.

Sockdolager: Booth threw open the door and aimed at Lincoln's head and triggered ancient laws of motion, the same laws that tossed lava blobs from volcanoes and tossed hellish waves in Sockdolager Rapid and kept planet bullets flying onward for billions of years. A pellet of metal went flying, knowing nothing of where it had come from and where it was going, knowing nothing of Booth's purpose and Lincoln's brain, knowing them only as a sudden acceleration and deceleration, tearing through Lincoln's brain knowing nothing of the values and purposes it was erasing, values that guided not just one man but two million soldiers who swarmed for four years to decide if human life has value, to decide if indeed all men are created equal; tearing Lincoln's brain into mere blood dropping to the ground, dropping like the Colorado River, returning Lincoln's brain to the blind, stupid universe from which life had worked so hard to emerge. An instant later John Wilkes Booth was dropping as dumbly as Lincoln's blood, falling at the command of the same gravity that smashed dumb asteroids on dumb dinosaurs, and he landed wrong and broke a bone, yet his cells were crying out for life and propelled him through the streets and countryside and woods, just like the cells of escaped slaves fleeing through the woods and swamps and not requiring the voice of a president to tell them their lives had value, not even the voice of God, for the cells were the real god and only pretended to be God and impersonated his voice in church. Animals heard the message that their lives had value, but this came with no detailed instructions and so humans argued and fought about it endlessly. Individuals and groups interpreted this message as applying only to themselves, and thus tribal gods assigned far greater value to their own people than to outsiders. Even when one omnipotent god declared he had created everyone and the same golden rules for everyone, these rules were often weaker than the genetic rules of ant societies, and humans continued philosophizing and fighting over the value of human lives. One three-pound lump of willpower declared that millions of people should no longer be valued solely as profits and expenses on plantation ledgers, should no longer have their lives commanded by whips, but millions of bodies collided over this, and now his will was invaded by an alien will and the value of millions of lives was flickering out, the entropy of the universe was lurching forward, even as plantation owners and slaves alike sat in wooden pews visibly wearing down and lifted Bibles visibly wearing thin and lifted voices noticeably getting older and sang hymns to the same god for promising to defend humans against entropy.

John Wilkes Booth sought his value by pretending to be people greater than himself, English kings and Caesar and Macbeth, but still he yearned for an important role lasting longer than two hours. He judged that the eleven-foot jump from the presidential box onto the stage was hardly worse than the jump he made onto the stage in *Macbeth* (a jump the critics dismissed as phony melodrama). Now, even as Booth proclaimed his immortal importance with the words with which Brutus assassinated Caesar, the ghost of Macbeth was standing there mocking him: *Life's but a walking shadow, a poor player, that struts and frets his hour upon the stage and then is heard no more: it is a tale told by an idiot, full of sound and fury, signifying nothing.*

Abraham Lincoln had imagined his body to be his, and it was true he'd been able to make it walk and wave and speak, but it had always belonged to forces larger than himself, to a nation of indifferent peasant cells living their own lives without knowing whether they were serving a president or a pauper, and now the body that had commanded two million soldiers was powerless to command itself and required other legs to carry it to his deathbed and grave. A century later, presidents would hold the power to ignite thousands of suns all over Earth and ruin billions of years of evolution and destroy animal and plant nations they knew nothing about and which knew nothing about them, but this would grant them no further power over their own life and death, which would be decided by the laws and outlaws of cells.

My ranger friend told me that when the massive search for Ron found nothing, the search leaders decided that Ron had indeed committed suicide and was "in the river." She offered no further

THE ABYSS

explanation for saying "in the river," and I didn't ask, but I did think of Ron's hidden shrine where he made offerings to the canyon. Perhaps in the end he had made an offering of himself.

In the river. I thought of the person I had met, now lodged among boulders at the bottom of the river, his head nodding constantly, his hair waving like algae, his arm waving at the vice president passing above him. I felt a sense of loss stronger than you'd expect for someone I'd met only briefly a year before.

I thought of how earlier this year I was heading on a Grand Canyon river trip and the river ranger put us on alert for someone who had drowned and disappeared in the river a few weeks earlier and was "nearly due" to reappear. The ranger explained the predictable stages of drowned bodies. As lungs fill with water, a body grows heavier and sinks to the bottom. It may roll along the riverbed for a bit but usually gets stuck in boulders. Within thirty minutes the skin turns waxy and blue, later greenish. Faces swell and become unrecognizable. Within hours rigor mortis sets in. Skin wrinkles and peels off and erupts in blisters. Fish have a feast, starting with softer tissues like eyes. As the body decays, it fills with gases and bloats, making it buoyant enough to rise to the surface. This process takes about six weeks, at least on a cold river like the Colorado. The ranger recommended several eddies about ten miles downstream from where the person drowned, eddies skilled at catching debris. We did look, but another trip found the body first.

Ron should reappear in a reliable six weeks. I counted the weeks until my next river trip. Six weeks. I was on the same schedule as Ron and might meet him again. I would be kayaking and freer than rafts to roam the river and eddies. From the start, I knew it would be not just a body I was searching for.

My upcoming river trip was organized by a Colorado-based kayaking school for which I'd worked for a couple of summers. While most kayaking schools prefer to start students in antiseptic swimming pools or calm lakes, our company owner, Bill, felt that kayaking was all about wilderness adventure and that the best way to get people hooked on kayaking was to take them on a true wilderness adventure, a week and eighty-nine miles in the Green River's Desolation

Canyon, offering some of the same grandeur as the Grand Canyon. The trip started out with two days of flatwater, giving beginners a chance to master basic skills, and then the rapids started out mild and stepped to more serious. My role was to take up the rear and rescue any kayakers who flipped and bailed out.

Rescuing people in the Grand Canyon is more demanding, for few rivers have waves so big and currents so powerful. The eddy lines below rapids are so turbulent that a swimmer trying to reach shore can get spun helplessly. Yet Grand Canyon rapids aren't as dangerous as mountain rivers choked with rocks and holes and fallen trees. Most Grand Canyon rapids—under normal conditions—are broad chutes of waves, their rare boulders and holes fairly easy to avoid. A swimmer gets swept right through the waves, feeling overwhelmed, but coming out safely. If a kayaker or swimmer plunges into a Grand Canyon hole, they may suffer the most violent pounding they've ever had, but the current should soon flush them out, though possibly with a dislocated shoulder.

A couple of weeks before our trip I learned that it would not be occurring under normal conditions. As part of the scientific study of the downstream impacts of Glen Canyon Dam, the dam was going to be holding the river at an unusually low level, 5,000 cfs, for days. I compared this schedule with our schedule and realized that the low flows would catch us just as we entered the Vishnu Schist gorge, which holds the most difficult rapids. Some rapids become easier as the river drops, huge wave trains becoming mild wave trains. But other rapids become more dangerous, a wave highway becoming a maze of boulders and holes. When we got to Sockdolager, I couldn't even recognize it. At higher flows Sockdolager holds the largest wave most canyon newcomers have ever seen, arching ten or fifteen feet and crashing like ocean surf, but now this wave was gone, replaced by two boulders with sharp holes. Sockdolager offered an easy detour around its obstacles, but other rapids would combine the worst aspects of Grand Canyon rapids and mountain rivers: a powerful current racing through tight obstructions, even forcing you to charge straight into holes.

Just when I'd gotten to know the river fairly well and was feeling comfortable with it, my map and confidence were being ripped away. I wasn't sure about my ability to take care of myself on this river, and the prospect of shepherding and rescuing others filled me with dread. But, I reassured myself, I was merely tagging along; the main responsibility fell to Bill, the company owner.

A few days later the office informed me that Bill was recovering from back surgery and could not go. That left me.

The day before launch I met the kayakers. Several were practically beginners. One was having trouble with her kayak roll (she would swim a dozen times on our trip). My heart sank. It occurred to me that Ron's body might not be the only body I'd end up looking for.

Fortunately, another kayaker showed up. Bill's kayakers were accompanying a raft trip run by a Grand Canyon rafting company, and that company's owner was giving a free trip to his good buddy Roy, with the understanding that Roy would earn it by looking out for the other kayakers. Roy was not just an expert but a daredevil, doing runs I would not dare. But sometimes the river would force all of us to be daredevils.

It would be a good trip for contemplating the value of human life.

THE PRIEST

Four years later, I spent two weeks on the river with another canyon lover who would later commit suicide there.

This trip was my private permit trip we'd begun by asking to be "blessed" by the Hopi salt pilgrimage rock art, at least as blessed as secular-minded people could get. I spent this trip riding on the raft of Dale, the chief interpretive ranger on the canyon's North Rim. He was highly adventurous, with years of rappelling into canyon caves, climbing mesas, and traveling the world in his off-season. He had a strong conscience: when a new park superintendent instructed rangers to stop talking about environmental controversies, Dale ignored him. Dale was endlessly learning and could talk about art, history, books, and philosophy. Some people would call Dale a humanist, but when you are measuring humans against the Grand Canyon every day, including tourist foolishness, you are not so tempted to think man "the measure of all things."

For our trip, too, Dale was being adventurous. He bought a new raft, a cataraft (with two pontoons instead of a ring), and this was its maiden voyage. Most rafters would hesitate to take an unfamiliar raft on a Grand Canyon trip and would give it a test run. Dale had done a lot of rafting but never the canyon. In boulder-thick Hance Rapid, Dale smacked an oar against a rock and the too-loose oarlock rotated, leaving the oar vertical, now a rudder steering us zigzag, so Dale had to finish some tricky maneuvering with only one oar. This improvising was typical of his high competence. Yet I'd also sensed a vulnerability in him, and when we got a bad surprise in Horn Creek Rapid, this vulnerability was ripped open, some old, deep wound.

Dale was a good fit for the role of park ranger, which works best with people with a love of exploring, both intellectually and physically. Being an interpretive ranger is an extrovert job, but it attracts a lot of introverts who as children drifted away from sports and into the woods or books, and it appoints them the spokespersons for world-famous natural wonders, offering some of the satisfaction a priest gets from being the ambassador of God, and offering a lot of admiration from the public.

Within a year of our river trip Dale had dropped into a depression that made it hard for him to do his job. Even everyday tasks became baffling. Having to back away from his job only worsened his depression. Did this loss of his social identity lead him to suicide?

Or was it a loss of nature? On our trip Dale was immersed in nature, happy to be seeing new dimensions of the place he loved, and he shared lots of natural history. He did enjoy his role as the conduit through which the canyon gets to speak. Yet if he was hearing better than anyone the canyon speaking, was the canyon failing to tell him that his life had value? Was it not his mind but nature itself telling him that he might as well die?

Life had billions of years to perfect its physical systems and mental programming for survival. As far as we know, other species don't give up on living and kill themselves. Life does self-destruct from cancer and other diseases, against all safeguards. Is depression just a mental cancer, suicide just an outbreak of blind disease without any legitimate message for us? Or did evolution blunder when it made humans, made minds so complicated and unstable that they tend to break down and override survival imperatives? Or perhaps the fatal error was that blind nature, a question that had never questioned itself, finally made a creature who was not blind, who could see the entire story of which it was a part, a noir story of creatures struggling and suffering and murdering other creatures and getting cancer and dying—a creature who finally had the vision and the nerve to ask the point of all this and saw only stupidity. And dreamed of meaning.

In trying to work his way out of depression Dale took an interest in paranormal things that used to be foreign to him. He seemed to be saying that the natural world wasn't good enough. For example, he took a keen interest in photographs from Mars showing a landmark that looked like a human face. The old Dale would have pondered this geological curiosity and its tricks of sunlight and shadows, but the new Dale was eager to believe the Mars face was the relic of a lost, spiritual civilization. I saw the psychology at work. Every day, Dale stared at huge cliffs and the cliffs stared into him, cliffs rough and collapsing, cliffs oblivious and silent, cliffs full of death, cliffs saying that humans are only a blip in the universe, and he wanted to see that the universe has a human face, even an angelic face. Not long before his suicide, Dale's mother died; the universe's nurturing face disappeared.

Dale's suicide hit me as not just the loss of a friend but, maybe, some sort of judgment from someone with high authority, a judgment I was left to try to figure out.

THE COYOTE

I had another friend who was thinking of killing himself at the canyon. Unlike Dale, who kept his decision a secret, Milo talked about it frequently.

Our faces flickered red from the campfire, and flickered a tiny bit more from the stars on this clear and moonless night. A few dozen feet away was a night without stars, the canyon. A Navajo woman set down a plate of fry bread, and we picked out fry bread and steaks and salad. Another Navajo woman spoke Navajo to her daughter.

When Navajos look at the night sky they see a distinctly Navajo sky, with constellations made of Navajo gods and heroes, with a mixture of order that upholds human life and disorder that mocks it. At the time of creation, Black God began taking crystals out of his fawnskin pouch and placing them in the sky in careful patterns that instructed people in how to live. The constellation Gah Heet'e'ii, or Rabbit Tracks, tells Navajos when it's the right season to hunt deer and teaches gratitude for the animals that give life to humans. When Black God sat down to rest, his star pouch still half-full, Coyote the trickster sneaked up and snatched the pouch and flung it into the sky, inflicting chaos on Black God's patterns.

One night around our campfire, under Coyote's sky, a coyote started howling, and another coyote answered. They howled back and forth, worrying rabbits, telling them that this is coyote's universe. Our little dog Poco lifted his head and listened attentively, and I wondered about his loyalties, whether the coyote within him was awakening.

One of the Navajo women commented, warily, that she had sold a big Navajo sandpainting today. Navajo sandpaintings originated as part of sacred healing ceremonies, aligning ill people with the order of the sky, and some Navajos feel it's wrong to sell sandpaintings to tourists.

We were outside Hermit's Rest, the rustic stone gift shop at the end of the eight-mile gated-off Hermit Road. Today the only hermit here was the shop manager, Milo—short for Milosevich—whose house was near the shop. The gate and the miles limited his social life, so sometimes he invited his employees to linger after work, for a cookout. I was here because I was friends with one of Milo's crew, Dave, with whom I'd done a kayak trip down the San Juan River. When we'd stopped at a huge Ancestral Puebloan rock art panel, Dave was ecstatic, for he was fascinated by Native culture. In the shop he enjoyed telling customers about the motifs and history in Native jewelry, rugs, pottery, and sandpaintings. Dave became such friends with Yoli, one of the Navajo employees, that when he went on to become a ranger at Canaveral National Seashore, Yoli, with no concept of how huge America is, how long a ride it was to Florida, got on a bus. When she arrived, she was bewildered as Dave explained that the rockets at Cape Canaveral weren't the same as the rockets at Disney World.

The campfire conversation meandered, and when dinner was done the alcohol came out. At this point Yoli always departed, and I usually did too, with some transparent excuse. If you stayed you had to watch friends becoming too drunk to walk straight, although campfire folklore claimed that Poco the dog went into sheepherding mode and made sure no one walked into a tree or over the rim. Theologically, this implied that we lived in Coyote's universe, with Coyote's cousin our only guardian angel. If you stayed you had to watch alcohol fueling Milo's despair, his lamenting that given all he had been through, for all he had lost, for what little he had left to hope for, for how little anyone cared about him, he might as well get into his car and drive to Hopi Point and into the canyon. One night Milo's threat so unnerved a newer employee, who was sure he meant it, that she camped out all night to guard his car.

On one visit, before dinner, I was sitting in front of Milo's desk in his tiny office in the back of the shop. Milo was checking over lists and balance sheets. From behind his desk came a sweeping noise, then a sob. Milo looked over his shoulder and smiled. The sweeping noise moved to the side of the desk. Milo looked over the side of the desk and smiled. Around the edge of the desk appeared a red bandana. The bandana was to cover scars and oozing flesh.

At his son's birth, Milo had watched in horror as his baby's skin began peeling off "like the skin of an onion," starting at the top of his head and spreading down his forehead and far down his back. Milo watched his baby convulsing with a stroke that would leave the right side of his body impaired, for life. The baby's cries were weird, partly from the cleft palate that would leave him unable to talk. His eyes were sealed shut with skin. With no epidermis to protect him from infection, he had to be rushed to a distant hospital and be treated like a burn victim. For two months he lived inside a plastic bubble, without human touch, with only a blanket for comfort, being fed through a rubber nipple.

Nathan peered around the desk at me, his eyes surrounded with

CANYON AND COSMOS

red scabs. He clutched his blanket to his chest. Nathan was nicknamed "Scooter" because his only way of moving was sitting on his rear and pushing with his good leg and arm. Milo joked about him being the official shop-floor duster: his sweatpants were continually dirty. He was now three years old but unusually small. He lacked some teeth. Chronic earaches had damaged his hearing. He could manage to stand up, but if anyone assisted him to stay up for too long he started screaming. For his first year his skin was peeling and oozing so constantly that he got glued to his sheets; now his skin was more solid but still peeling and itching, varying unpredictably every day. He had no sweat glands, leaving him more vulnerable to heat.

The doctors diagnosed that Nathan had a very rare disorder, always fatal. He would be lucky to live to age twenty. Nathan would have to go through all the usual struggles of growing up, plus many more, but never enjoy the rewards. Milo and his wife Monica might have to devote twenty exhausting years to a hopeless cause. Twenty years of torturous emotions. Are you supposed to feel guilty for inflicting all this suffering on an innocent child? Or can you resent your baby for inflicting this horror on you? Can you blame it all on someone's bad genes or bad mutation-inducing habits? Of the parents giving birth to babies with this disorder, 80 percent of their marriages collapse.

What did science have to say? Scientists could talk about genes, disorders, statistics, and even treatments, but when it came to why the universe is imperfect and inflicts pointless horrors on its creatures, scientists can be as helpless and evasive as anyone, including theologians. Milo did latch on to one theory. His house was built atop rocks brought in to minimize the sloping land, and Milo decided that these rocks were uranium ore from the Lost Orphan Mine a few miles down the road, which once yielded some of the richest uranium in the country and reportedly caused the deaths of some workers from cancer. If Milo and Monica and their fetus had been living atop a radioactive dump, this explained everything, removing personal blame, making it a fault of technology, preventable through better management. But Milo's uranium theory was wrong—I checked it out. Milo also mused that if there was a God, God was grotesquely

THE ABYSS

incompetent, if not a malicious psychopath. Yet then Milo said that God had sent Nathan to him to make him a more loving person, and that he wouldn't trade Nathan for a healthy baby.

Monica drifted toward the idea of reincarnation, which gives logic and justice to suffering. There is no such thing as undeserved suffering. Nathan was scooting on a trajectory he had set in motion in a past life, receiving terrible justice for what must have been a terrible crime. What crime would justify a baby's endlessly burning skin? Nathan must have burned the skin of innocent children. Nathan must have been an SS officer at Auschwitz, in charge of funneling trainloads of innocent children into the gas chambers and the ovens where their skin peeled off from head to foot and oozed out blood and juices. But what kind of religion is this, encouraging mothers to look on their babies as monsters?

The Jews—and the Christians too—Nathan is herding off the trains have a different answer. There is indeed undeserved suffering, lots of it, but this was never God's intention. God is full of justice and compassion and he designed the universe to run on justice and compassion. This idea was once radical and powerful, that you might be protected not by gangs of imperfect, squabbling gods but by the omnipotent benevolence that controls the universe. But this idea rested on hopeless contradictions, leaving God with the blame for evil, forcing theologians to shift the blame onto Adam and Eve. But this idea, too, was full of theological landmines. So God came to Earth and got crucified and gave undeserved suffering the most wretched face and the highest importance it ever had, and everyone could know that God now felt complete personal sympathy for their own undeserved suffering, even if beleaguered priests and rabbis admitted that in the end it was all a mystery.

To the Jews Nathan was herding off the trains, it was all a mystery why the just God who rules the universe had abandoned them, yet again. The rabbis watching the black smoke roiling out of the crematorium chimneys saw Roman torches burning Jerusalem. The carbon-black smoke roiled up all day, twirling like swastikas, and dissipated into the sky; the night-black smoke roiled up all night, revealed by the column of stars wavering in and out of eclipse. This carbon originally came from the stars and was trying to return to the stars, and in between it had imagined that the universe is ruled by a compassionate judge, but it had gotten into endless arguments with brother carbon who held that the universe is ruled by reincarnation or Coyote or brute force, and today at least brute force had triumphed and was obliterating the foolish mistake of cosmic compassion. This carbon was born from gas chamber stars and crematorium stars and ash-scattering stars and now it was being returned to where it belonged. This is the way a universe of force behaves when it evolves from supernovae to volcanoes to hungry, fighting, bloody animals to humans with philosophies. The three-year-old Jewish children Nathan is herding off the trains look at him in his black SS uniform, black with various sparkles, and they imagine Nathan is the golem of the night, the night itself given human form to walk on the earth, just like Frankenstein's monster, just like Jesus, the night proclaiming that the glorious galaxies praised by rabbis and priests and astronomers and poets alike were nothing but swirling swastikas.

Milo lifted Nathan off the floor and set him on the desk. Milo adjusted the bandana to hide the red scabs, probably for my sake. For a moment Nathan was calm, and then for no obvious reason he started sobbing. Milo gently asked him what was wrong. Nathan continued sobbing, clutching his blanket. Milo said that Nathan had good days and bad days, and you could never tell. Nathan dropped his blanket, which snaked and went over the edge of the desk. I reached down and picked it up and tried to dust it off and handed it back to him, to his one working hand. He clutched it for a moment and then not only let go of it but tossed it toward the edge. It fell to the floor. I picked it up again and held it not to his hand but to his chest. He had stopped sobbing. He grabbed the blanket only long enough to toss it to the floor again. He smiled. Here in a world of bodies coming and going and acting inexplicably, bodies with powers he kept trying and failing to evoke from his own even more inexplicable body, here was a body his will could operate, evoking it to reach out farther than he could and reliably return to him what had long seemed the only comfort and security the world had to offer.

From the Journal of Tiki, a Youth Dying Young

Left in a secret spot in the canyon as a memorial:

> *Sometimes as I lie here at night I like to look around my room because at night the lights play shadows on things and there are so many different things which have so many different stories. Over on the desk I have a candle and it's glowing red and it gives off a beautiful shimmer to the room. And just behind it is a Spanish Madonna. . . . The Madonna, because of her gentleness, seems to watch over me and say, "Don't worry about anything—everything will be alright."*
>
> *I wonder if it is true that if you say "I love you" to my stuffed animal, then he is real.*
>
> *"No," the candle says.*
>
> *Now the candle gives out a little—it is not quite so bright.*
>
> *Man searches so far for answers. . . . Why is it that man always wants someone to give him answers?*

What the Thunder Said

Out of the secret tectonics of the sky, out of the collision of air continents, out of the blue, cloud mountains materialized and rose higher and spread wider and grew darker, projecting their darkness across the ground. The clouds began flickering and rumbling, the latest, local outbreak of the thunder that has roared continuously somewhere on Earth for billions of years. If thunder is Earth's strongest and most ancient voice, what is it telling us? Is it the voice of rain granting us life? Or is it the voice of chaos?

In a swifter version of the erosion by which mountains disappear grain by grain, the cloud mountains began eroding.

We watched the storm gathering ahead of us, moving toward us. The sky grew eerie dark. The wind went crazy, rocking trees wildly, sending waves rolling upstream, tossing spray into our faces, nearly flipping our kayaks. An eclipse curtain rushed toward us; rain engulfed us. My kayak became a drum being played by the sky, by the same vast music in which my heart too was a drum. Thunder roared. From the cliffs high above us, waterfalls began breaking out, dark brown, some dropping a thousand feet.

As far as one hundred miles upstream, this storm overflowed creekbeds, bulldozing boulders into the river, constricting rapids, creating new rapids and riffles.

This thunderstorm would last for centuries. Grand Canyon rapids are the ghosts of thunderstorms. Rapids lie at the mouths of side canyons, down which flash floods swept debris into the river, forcing the river to plunge through it. Some rapids gradually fade away, but then new storms rebuild them, maybe worse. One rapid might be the composite of hundreds of storms over thousands of years. Throughout the canyon, ancient thunder still echoes, ancient violence shakes logs and boats and people.

To the thunder they would come. On the path of John Wesley Powell, who lost his arm to Civil War thunder but was determined to go on living, they would follow. Some came because they too were wounded, if only by being born human. They came to weigh their lives on a scale they couldn't find elsewhere, and some died, yet some found life. To a canyon full of ghosts, they came.

Snow was falling yet again. Once, so long ago, he had loved snow, the beauty of it, the mountains of it, the skiing down it with speed and grace. Yet now the thickening snow was a cold prison wall, reinforcing the barbed-wire fences and guard towers and guns and endless Russian plains. The wind cut through his meager tattered coat and flushed him with cold, reinforcing his hunger and weakness and his feelings of being lost, hopeless, and worthless. Everything he had once imagined himself to be had been crushed out of him, including his pride in being a soldier standing with long rows of heroic soldiers, with red flags and black swastikas saluting them, with their god telling them they were stronger and nobler than all others. Their pathologically insecure god, needing to put down others, deeply down, millions down, to lift himself.

It had all been an illusion, all crushed.

Gazing through the snow and barbed wire, he saw something very far away. He saw mountains of snow melting and turning into beautiful rivers running through beautiful forests colorful with springtime, rivers tumbling through rocks, and on the river he saw a kayak as alive and beautiful as a wild animal, and in the kayak he saw himself, racing through the waves and rocks, in total control of his fate, flying with total freedom.

Had it really been himself in that kayak? Could that ever be himself again? And if he ever got out of prison, could he ever get the prison out of himself?

"He was only fifteen when he was drafted into the German army," Fletcher explained about his friend and mentor Walter Kirschbaum. "It was toward the end of the war and the Germans were taking younger and younger men. He believed in the Nazi cause in the way that any fifteen-year-old believes in an idea he's been fervently taught his whole life."

The noise of a small riffle blurred Fletcher's words about Kirschbaum fighting on the Eastern Front, a hell of scorched earth and brutality. I paddled faster to keep up with Fletcher, not always easy as he was the many-time national champion at kayak racing and now the coach of the US Olympic kayaking team. "The Russians held him for several years after the end of the war. When they finally released him, he'd spent a large part of his life imprisoned. He was haunted by it." He was emotionally crippled by it. Several times, Fletcher emphasized Kirschbaum's severe lack of self-esteem. "Clearly nonsense." Kirschbaum was the world champion kayak racer, the first to kayak the whole Grand Canyon and Cataract Canyon, the first to make many treacherous first descents in Europe and America, one of those with Fletcher. Yet Kirschbaum always acted as if this was nothing, as if it was just an accident he was world champion. "Clearly nonsense. No one is world champion by chance." When Fletcher met Kirschbaum, "the man had just beaten Milo Duffek to win the world championship." Fletcher told of how the Czech Communist government wouldn't let Duffek travel to the West for races without a guard. In one race, Duffek charged across the finish line and never stopped, disappearing around the bend to his getaway car. Duffek was free.

For Walter Kirschbaum, too, kayaking brought freedom.

We were heading into Westwater Canyon in Utah, a short exposure of the same schist and granite exposed at the bottom of the Grand Canyon. Here, too, the schist pinches the Colorado River into rapids with big drops, big waves, big holes, and strong eddies, plus one of the most notorious whirlpools on the entire river, the Room of Doom. In 1962, with three friends, Walter Kirschbaum made the first kayak trip through Westwater Canyon. Before my first Grand Canyon kayak trip I'd come here for better training, and for each subsequent Grand Canyon trip I'd come back for a good warm-up. With another Grand Canyon trip coming up, I'd gotten a Westwater permit but needed someone to go with, so I called up Fletcher Anderson and his wife Ann Hopkinson, who had their own Grand Canyon trip coming soon, and asked them if they, too, needed a warm-up. The night before our trip I camped beside the launch ramp and was pounded by a ferocious thunderstorm, which brought down tree branches. In the morning, the river had turned black. A recent mesa-top forest fire had left many miles of ash and burnt trees, and the storm had flushed it down. I rather liked the idea of kayaking on a black river, but the prospect of being rammed by a log while running a rapid was not so charming.

Fletcher's father, Keith, was a pioneer of Colorado whitewater and helped start an annual race on the Arkansas River at Salida that soon became America's premier race—Ann won the women's division several times. When European kayakers came to Salida in the early 1950s, they left the Americans far behind, but they learned a few things. They were still using wood-framed, canvas kayaks, while Keith Anderson was experimenting with fiberglass. From Keith, with young Fletcher looking on, Walter Kirschbaum learned to make fiberglass kayaks, though he remained wary of their weight, maybe seventy-five pounds; so for the kayak he built for the Grand Canyon he compromised: a fiberglass hull and canvas deck, maybe forty pounds. Far more importantly, he discovered that winning races was not important compared with the grandeurs of wild canyons and the almost mystical feelings they inspired.

But that didn't happen, Fletcher explained as we continually painted our paddles black, until the Grand Canyon. Until then Walter was obsessed with racing. It was good therapy. He had emerged from prison with chronic anxiety. On an airplane, at the mercy of inexplicable machinery and an unknown pilot, he visibly trembled with fear. While in his kayak, even amid deadly obstacles, he felt in control. Prison also left him with a void of identity, and winning races gave him a sense of worth. But never enough, never for long, so he was driven to continue racing and winning. After years of malnutrition he never recovered his physical strength, but he won races on his technical strengths, from flawless maneuvering. Walter also tried to escape his demons through alcohol, but this didn't work nearly as well as kayaking.

I was reminded of the other Walter who emerged from a POW camp raw and broken. In the Civil War, John Wesley Powell's brother Walter was captured and spent six months in a Confederate stockade, in horrible conditions. Walter lost his physique, his health, and his mind, becoming morose and erratic and angry. John took Walter on his Colorado River expedition hoping it would help cure him. Walter did see the beauty of the river and canyons, but he remained troubled and troubling, and eventually ended up in a mental asylum.

Walter Kirschbaum had something else to escape. One day in Germany he'd been driving down the street when, from between two parked cars, a little girl ran into the street, and Walter hit her and killed her. He had been drinking, but not much. The police did not blame him. But Walter blamed himself. And the German newspapers couldn't resist a celebrity scandal: *World Champion Kayaker, Alcoholic, Kills Little Girl*. Now all Germans looked down on him as a killer, the same Germans who murdered a million little girls without regret. In America, no one would know this story, and he could define himself through his actions.

No, Walter hadn't hit her, his car hit her, a metal embodiment of the same dumb Force that raced unstoppably as rivers, far beyond human control. The little girl lying crushed in the street was only the latest proof he had seen of the weakness of human bodies compared with Force. Many times he had been called out to search for people drowned in rivers, bodies broken and rotting, crammed into logjams or floating along. Even one bottle of water was stronger than a man, if it was sprinkled with grape guts. No, even though every part of a car had been designed and constructed for human control, his car and every Panzer tank was secretly controlled by primordial, mindless Force. Yet in his kayak he confronted Force and Chaos and he outwitted them, he was in control, he was exhilarated by his control. Boulders never moved, and waves were more predictable than humans. A river allowed you to make judgments and choices and actions and it enacted them with its own power, magnified them into larger-than-life motions, ratifying you with a verdict as decisive as the difference between life and death. In his kayak he hadn't killed anyone, not even himself, though people warned him that trying to kayak the Grand Canyon was suicide.

It's interesting, said Fletcher, that Walter with all his experience and skills had to go through the same learning process as every kayaker in the Grand Canyon, at first underestimating the scale and power and push of the waves. Even seventy-seven miles into the canyon, Walter made a major misjudgment. He carefully scouted Hance Rapid, one of the worst obstacle courses in the canyon, and imagined he would enter far right and cross its entire width to end up far left, but the current, at 40,000 cfs, swept him far off course. A wave slammed him over. He tried to roll "but failed due to the sweeping power of these haystacks," he would write. "A giant's fist, then, it seemed, dragged me out of my kayak." He struggled to get to his kayak and to shore, but "the whirlpools never seemed to end, and when one of them released me another was waiting to take over." At last he reached shore, but there were no longer beaches, only weird black cliffs.

He was entering the Vishnu Schist, hard and dark, polished into surrealistic spires, streaked with granite, full of shadows. John Wesley Powell entered the schist with dread: "We can see but a little way into the granite gorge, but it looks threatening. . . . It inspires awe. The canyon is narrower than we have ever before seen it; the water is swifter."

For his own amusement, Vishnu incarnated himself as a human child, and one day his human mother gazed into his mouth and beheld the whole universe. Now, as Vishnu and his charioteer Arjuna stood waiting for battle, Vishnu revealed his true identity, a radiant being containing every star and god, with infinite shapes and faces. If the

THE ABYSS

radiance of a thousand suns were to burst forth at once in the sky, that would be like the splendor of the Mighty One. Arjuna beheld a vision of the destiny and ultimate significance of all earthly life: "As the torrents of many rivers rush toward the ocean, so do the heroes of the mortal world rush into Thy flaming mouths. As moths rush swiftly into a blazing fire to perish there, even so do these creatures swiftly rush into Thy mouths to their own destruction."

John Wesley Powell and Walter Kirschbaum rushed into the mouth of Vishnu.

For eons, the Vishnu Schist lay hidden underground. It knew nothing of the changes happening above it, the order of events that would lead it to being called Vishnu. Then the schist was unburied and felt sunlight again. Rushing water carved it into odd new shapes, and one day rushing water carried into it more odd shapes, floating things, wounded things, a man with only one arm, a man clinging to his overturned kayak, both whirling in whirlpools, frightened, looking at the rising black cliffs with dread.

During the schist's eons underground, life's groping for identity had turned into a deep dreaming, insecure little creatures dreaming of being important, fragile creatures dreaming of being immortal, and to persuade themselves those dreams were true, armies swarmed over the earth fighting armies that held different, contradictory dreams. John Wesley Powell and Walter Kirschbaum were refugees from that swarming. As they entered the schist, as they were swallowed by the mouth of Vishnu, they were meeting, across eons of transformations, the rock-remembered beginnings of life's quest for identity, the beginning of those swarming armies. They looked at the schist with horror. Was this horror the correct verdict for life to pass on the journey it had made? Was the emergence of humans the triumph of the universe's order-making genius? Or had evolution, in its blind groping, incarnated its blindness and chaos into human form, producing utter madness?

In the night, in the empty New Mexico desert night, Robert Oppenheimer paced nervously, stared into the darkness, and put a cigarette into his mouth, twenty-two years before he would die of throat cancer. A small flame erupted, Oppenheimer's face flickered for a moment, and disappeared.

Around him Oppenheimer heard tense and quiet talking, in accents from all over Europe. These men had fled their homes to escape from Walter Kirschbaum. Some of these men belonged to a maltreated desert tribe that had discovered itself to be God's instrument for proclaiming that the universe is ruled by order, love, and justice. They had fled Walter Kirschbaum because Kirschbaum believed that the universe is actually ruled by Force. Or thus Kirschbaum had been told by a man who, walking home from the opera one night after seeing Siegfried triumph heroically over the dragon, could no longer bear to be a weak, insecure, anonymous man alone in a shabby room and decided he was the prophet of the Master Race. They had fled Kirschbaum because in order for Kirschbaum to play Siegfried, someone had to play the dragon. They had come to the New Mexico desert because they were afraid Kirschbaum would prove that the universe is indeed ruled by nothing but Force.

A warning siren sounded, and the men took cover in trenches. From loudspeakers came a voice: *Ten, nine, eight*—it would all add up to zero. *Three, two, one, ZERO!*

And there was light.

Without a sound, a point of light erupted and swelled into a brilliant halo. Around it, far around, the desert sands shimmered eerily. The halo became a brilliant red fireball, too bright for human eyes. *If the radiance of a thousand suns were to burst forth at once . . .*

It was the glow of ghosts. It was the death agonies of stars. It was the energy of supernovae, locked into uranium for billions of years and now uncaged. A supernova was raging on Earth. Far away in space, billions of light-years away, the energy of those supernovae was only faint ripples of light, moving ever farther away from Earth, yet here they had been summoned to their former ferocity, summoned because Walter Kirschbaum wanted to play Siegfried. *If the radiance of a thousand suns were to burst forth at once in the sky, that would be like the splendor of the Mighty One.* The fireball rose into the air, still silent. Around it the air glowed blue. Beneath it, dust and sand rose

THE ABYSS

into clouds and swirled crazily. The men felt the heat of supernovae on their faces. Abruptly the silence ended: thunder roared. Thunder shook them. A hot wind swept through them. Oppenheimer stared at the fireball with silence and dread and thought of the scene in the Bhagavad Gita in which Vishnu reveals his true identity and awesome power. As the thunder rumbled on, Oppenheimer remembered Vishnu's words: *I am become death, the shatterer of worlds.*

Before long, Earth would be covered with thousands of weapons far more powerful, made not from uranium forged in supernovae but from hydrogen forged in the Big Bang. In its cages the energy that had created the universe would wait, wait for the command to devour what it had created. All over Earth the Big Bang would rage. The fourteen-billion-year quest for order and identity that began with this fire would now be destroyed by it. The life that had arisen from the chaos of this fire would melt back into that chaos. Or perhaps life had never truly emerged from that chaos. Perhaps the Big Bang had given birth to more chaos than order. Perhaps in the wreckage left by the Big Bang, in the blackness adrift with cratered planets and failed stars and raging fires, life was a cancer growing toward self-destruction. Perhaps through the portion of its chaos it embodied in humans, the Big Bang expressed its own true identity by packaging itself as a world-destroying bomb. Perhaps in the black desert night, through a dread-filled human mind, the Big Bang finally found words for what it had announced with its first eruption of infinite light: *I am become death, the shatterer of worlds.*

Out of its night blackness, out of its disguise as mere rock, out of the flickering incarnations of two billion years, Vishnu opened its mouth and revealed itself as John Wesley Powell and Walter Kirschbaum being swept along. They had come here because they were full of chaos, because they were fleeing the chaos of war and the human chaos that led to war. They had come to this chaos to confront the chaos within themselves. In wave mirrors they saw themselves and the denial of themselves.

Powell compared his canyon journey, where the shadows magnified his own gloom, to the despair he'd felt when imprisoned in his hospital bed after the Battle of Shiloh, his arm gone. Emerging from the canyon gave him an enormous sense of triumph and peace and beauty: "What a vast expanse of constellations can be seen! The river flows by us in silent majesty . . . our joy is almost ecstasy."

It was not only psychologically or symbolically that Powell and Kirschbaum were confronting the chaos within themselves. The primordial chaos embodied in the Vishnu Schist had sent life on its journey to Shiloh, Auschwitz, and Hiroshima, and the river raging through the schist still contained the chaos from which life had never fully emerged. By confronting that river chaos they were facing a deeper chaos, submitting themselves to it, struggling through it, daring it to hurt them further, daring themselves to emerge from it, to win some kind of victory for life.

"It was only during his Grand Canyon trip," said Fletcher as our kayaks drew a common wake, "that Walter finally felt released from his past. He finally felt at peace with himself. He came out of it a different person." Fletcher didn't offer more explanation, and I didn't need to ask for one. We both knew plenty of people for whom a two-week immersion in the Grand Canyon had changed their basic sense of reality and of self, who emerged as truly new people.

I thought of Walter Kirschbaum surrounded by Grand Canyon cliffs, finally a barrier large enough to hide human society and its arrogance and wars and gulags. I saw him on the river, finally a force more overpowering than his demons, rapids louder than artillery and little girl screams. After his swim through Hance Rapid, which would have demoralized most kayakers and left them dreading the next rapid, Kirschbaum said he felt liberated. I saw him amid canyon sunsets and moonglow waters, finally a beauty strong enough to match the ugliness he carried. I saw his past and his grief draining out of him and into the river and sand, like into a Navajo sandpainting, and being washed away.

Walter, explained Fletcher, became enthralled by the canyon. It became very powerful for him. "The Grand Canyon changed Walter's attitude about kayaking. He had worked hard to become world champion, but after the canyon he lost interest in competitive racing."

Kirschbaum wrote: "I found wilderness canyons so attractive that all the success I had had in slalom and downriver races of national and international level faded in importance compared to the deep impulses experienced on my excursions.... Canyon expeditions gave far more and deeper satisfaction." Walter became obsessed with the Colorado River, with exploring every whitewater section of it and its major tributaries, some of which were remote and treacherous and long assumed to be unrunable. Walter was still a driven man, Fletcher admitted, but in a different way: it was only in the majesty and chaos of canyon rivers that he could feel at peace.

Fletcher knew all about being driven. A dozen years before our trip he'd been driven to kayak the Grand Canyon solo, in forty-nine hours, setting a new speed record, though his larger goal was to break out of the "vacant shambles" in his life. He had moved away from the mountains to a big city and was doing a job he didn't like for the sake of money, cutting himself off from everything he loved. Night after night he was haunted by the same nightmare of being in his kayak, being pounded by the monstrous hole in Lava Falls, and then magically rising into the air, out of the canyon, over the volcano on the rim, leaving him feeling "excessively peaceful."

As soon as he launched into the Grand Canyon, Fletcher felt "I was at last alive again," felt he was in "a holy place." He paddled at racing speed, making twenty-six miles in three hours. His tendonitis began acting up, then hurting badly. His wrists began swelling, forcing him to slow down. In Horn Creek and Crystal Rapids, he flipped. As he pushed onward in pain and exhaustion, the canyon became a blur and he began making sloppy mistakes. At Lava Falls, for the first time on this trip, he stopped to scout. In eight previous trips he had flipped in Lava Falls four times. This time he started out correctly, but his wrist pain led him to let up too soon and he fell into the very hole of his nightmares. He was punched violently and wanted to scream, but pulled himself out and up.

In the end, his speed record didn't feel important. All the way through the canyon he felt bad that he was trying so hard to "spend the least amount of time possible in the most fantastic place on Earth."

From a mile away, Walter heard chaos announcing itself with a deep rumbling. Walter saw the strata and colors of the cliffs disappearing behind curtains of lava that had flowed down from the rim. It was a violation of the canyon's order, verticality violating the horizontal, blackness hiding the red and white, roughness replacing limestone smoothness, fire replacing water, roar obliterating the peace he'd started feeling.

Some of Walter's kayaker friends had told him he would be foolish to try Lava Falls—there were some things a kayak was never designed to do. The Park Service had told him to portage it.

He landed and scrambled up the lava talus and there it was, spread out before him: chaos itself, roaring at him, roaring inside him. Chaos was a racing, churning, violent whiteness. A whiteness imprisoning him, like the blizzards of his Russian prison. A whiteness roiling like the smoke of Russian villages and Auschwitz and Hiroshima, convulsing like a little girl dying in the street.

Atop a volcanic ridge, atop the collision of continents and the grinding of rock into lava, atop the chaos of lava that had welled up from the earth and become a chaos of water, atop the thunderstorms still worrying their way downstream, there now stood a further stratum of chaos, welled up from the same chaos as the lava and water yet churning more strangely. He stared into the chaos of Lava Falls, comparing it with his own chaos. Which was stronger? Would his own chaos compel him to challenge Lava Falls? He stared at where the whole center of the river plunged into a monster hole he would call "horrifying . . . where a kayak would definitely be lost." He searched for routes that evaded this hole and saw trouble everywhere. He stood there, the conflict of water and lava become his own bewilderment. He stood where the war-wounded John Wesley Powell had stood nearly a century before, confronting chaos.

In allowing Lava Falls to symbolize the chaos of the universe, I am not suggesting that by merely bouncing through some waves a human could conquer or heal the chaos of the universe. Of course, when he reaches the bottom, the universe will not have changed. Stars will still explode, volcanoes will still erupt, animals will still devour one another, and humans will still be desperate to find significance.

THE ABYSS

Yet in so far as cosmic chaos knotted itself into human wars and the psychological chaos of their minds, Powell and Kirschbaum at least might unravel this apex of cosmic chaos. They had gone to a great deal of trouble to seek out chaos at its source, deep in nature, literally deep. They did not appear to agree with the psychologists who say that happiness consists of being well adjusted to human society. They might be haunted by existential dread, but they did not seem to agree with existentialists who say that humans are free to invent themselves in city cafés. Their canyon journeys were archetypal, a good fit for the quests of mythological heroes. Homer could have sent them on their monster-fighting odysseys. They were repeating the quest of Tiyo, the mythic Hopi hero who went down the Colorado River in his hollow log, not merely to run rapids, certainly not to feel he had conquered the river, but to reach deeper into the secrets of nature and dispel chaos and bring greater harmony to his people.

And what wisdom would Powell and Kirschbaum win?

If there was anywhere on this river that would convince them that nature is a madhouse, it was Lava Falls. Here it was harder to believe in the benevolent order of the universe. Here it was harder to remember why humans should inflict such chaos on themselves when it would be easier to quit.

And if there was anywhere on this river that would prompt them to find God, it was Lava Falls. Here the strongest human bodies were too weak and the strongest wills could falter and be tempted to rely on something far stronger than themselves. If a quiet and safe view from the canyon rim could inspire many people to see the work and feel the presence of God, how better is a cataract like those that humans everywhere filled with gods. John Wesley Powell had fled from his father's desire for John to follow him as a Methodist minister, fled into the new worlds of natural history, went into the Grand Canyon to find its meanings and his purposes. Yet when his exhaustion was lifted by a beautiful creek, he named it Bright Angel, as if Powell would not mind having the presence and support of angels, as if he were missing the comforting universe his father had promised him. God rebuked Job with less thunder than Lava Falls.

Chaos did interrogate Powell and Kirschbaum more thoroughly

163

than it did most people, and they saw it more deeply. It had interrogated them in battle and impaled itself into them as an aching lostness. Through the eyes of strangers it stared at them as worthlessness. In the canyon, chaos wrestled against every stroke and turn. In the shadow of chaos, God lurked, watching for an opening. Yet neither man yielded to chaos or God. Powell came to the canyon because he believed that the only grounds for reality and human life was the ground itself, and that the Grand Canyon offered Earth's deepest look into that ground, into the natural forces that gave birth to humans. In his summation of the canyon, Powell admitted it was a hard place for humans to inhabit but insisted it was Earth's ultimate grandeur and beauty, which should inhabit us. It was strange and demanding and unreliable, yet it was the only God we had. Thus spake Odysseus. Walter Kirschbaum wrestled with chaos in a less theological way, more alone and small, yet he found that chaos could have a sublime face, that he could converse with chaos, that if chaos in its depths did not care about human wills and troubles this might actually be forgiveness, a blessing, and this gave him peace.

Yet both chaos and God could be sneaky and persistent and vindictive. Which one was it that one day paid Walter a visit and disguised itself as a clear, calm pool of water? Some of this water could have been the same water he'd seen in Lava Falls years before. Now it was safe and servant, a bathtub full of water. Walter slipped into it, into his own image. From next to the bathtub he picked up a bottle and drank what could have been the same water he'd tasted as he made it through Lava Falls. It tasted better, more soothing. Walter had never been able to outrun this soothing. He took another drink. He raised his head and it hit the underside of a cabinet and it knocked him out. He slid into the water, too deeply. And he drowned.

Far ahead of us we heard thunder, the roar of Skull Rapid, named for a human skeleton found there. It filled me with dread. Skull had become my worst rapid anywhere, repeatedly tricking and outmatching me. Today the Room of Doom, energized by its socket in the cliff, was a solid yet churning mass of ashes and burned logs, a nightmare for any swimmer or kayaker who ended up there.

Fletcher ran without scouting and made it look easy. Ann and I scouted, and Ann ran it well. Once again Skull tricked me, surfing me too far right and through two holes, the second one flipping me. The current began knocking my head against the cliff, hard, repeatedly. I knew this was dangerous, risking a concussion or blacking out; I knew I should bail out immediately and start swimming away from the Room of Doom, but kayaker pride equates swimming with disgrace, and I waited and got knocked more, then bailed out.

Safely ashore, I was examined by nurse Ann for signs of a concussion; she pronounced me okay. Skull Rapid pronounced itself stronger than human skulls. The river pronounced that chaos had triumphed, yet again.

BLACKTAIL CANYON

Dale and all of us loved it here, at this place, on this Earth, under this full moon. We had made an effort to be here.

As sunset faded, we walked up the creekbed to where Blacktail Canyon narrows into a slot canyon. We carried guitars and voices. Blacktail Canyon offers the best acoustics in the Grand Canyon. Water was already whispering here, and frogs were singing their desire to continue to be frogs. We had planned our schedule to camp on the beach at the canyon mouth just so we could make music here. Dale had upheld the long tradition of rangers who at evening campfire programs got out a guitar and sang, leading a sing-along.

Blacktail Canyon holds the seam between the Vishnu Schist, sinuous from occasional flash floods, and a rock stratum more than a billion years younger, a gap once full of rock, thoroughly erased. Here you can hold your hand against the cliff and bridge more than a billion years. Perhaps you can feel those eons still alive within your own body. But the rock is also telling you that if a billion years of strong rock can disappear completely, a human body is far more fleeting.

Even if you don't know its geology, Blacktail Canyon has a surrealistic beauty, even in daylight, and we were seeing it as few people did, as blackness merging with the blackness of the sky, except that tonight the moon was full and moonlight was creeping down the cliff and toward us. It was magical.

Dale started playing and we started singing, songs we knew, needing no flashlight to read the words and disrupt the night. Some of our songs spoke of time passing, but only as humans losing youth or love or hopes. Still, the rocks had finally obtained a voice through which to speak of time passing, if only obliquely, eons of flows and sedimentation and erosion encoded as wistful regret. Off the Carnegie Hall cliffs the guitar notes and our voices resounded vividly. Off the geological cliffs they rang ever deeper. Blacktail Canyon was telling us that even erosion could be deeply beautiful. The frogs sang with us. The moon was coming. Long memories were being born, full of beauty, music, adventure, friendship—everything good. We didn't want to leave and lingered late. Our time here was already too short, and never again.

Why would anyone choose to leave this Earth?

Coyote Baffled

One time as I was driving to our campfire dinner at Hermit's Rest, something odd happened. A jackrabbit came racing down the road, right at me. I braked and stopped and the rabbit ran right under my truck. In my rearview mirror I didn't see her emerge. I looked ahead again and saw three coyotes braking and stopping. The lead coyote stared right at me, seemingly indignant at this odd trickster trick I was pulling on them. The coyotes continued staring under my truck—the rabbit was still there. The coyotes wanted to approach but were fearful, as if they had seen rabbit-chasing coyotes getting hit by cars, and perhaps this was the rabbit's strategy, but I had stopped and now the rabbit was trapped, and the coyotes were trapped, and I was trapped. Deciding what to do meant deciding the value of rabbit lives versus coyote lives. I sat there, baffled by this trickster trick the universe was playing on me.

My sympathies were with the rabbit, but could I justify this? Rabbits are gentle creatures, not bothering other animals, bringing Easter eggs, while coyotes terrify gentle animals and rip them apart. My sympathies came mainly from my childhood pet rabbit, who one night inexplicably got loose from her backyard cage and was killed by a dog. Thus I glared at the coyotes as if this ancient murder now had confessors, as if they were not even animals but mythological Coyote who had sown chaos into a world that could have been kinder. The coyotes looked at me with puzzlement, as if Coyote himself had no idea why animals ended up confronting one another on an empty road on the edge of an abyss, forcing decisions about who goes on living. I realized that to the coyotes my face might be the face of cruelty, forcing them to go hungry. Had the coyotes ever asked to be predators, with all its struggles? Coyotes hadn't introduced hunger into the universe. Coyotes were trapped in the same desires as rabbits and me. What god had made me the god to judge other animals and decide the value of their lives? Let us prey.

I thought of honking my horn to scare the coyotes away, but this might also terrify the rabbit. I thought of creeping forward, but then I might run over the rabbit. I didn't see the rabbit run off, but I saw the coyotes warily moving around me and continuing into the woods.

As Nathan and I played blanket fetch, Milo continued working on his lists. Above him were shelves packed with merchandise that translated the canyon into human shapes and identities, into coffee mugs and T-shirts, into status symbols by which Grand Canyon tourists could outrank lesser destinations. "Our society," Milo complained, "has trained us to approach everything as consumers. We haven't been to the canyon unless we've bought it."

One outside vendor brought in petrified wood items, appropriate for a geological park, but the one that most intrigued me was the plastic Jesus crucified on a petrified cross. When this wood was alive it was touched by dinosaur shadows and by small mammals fleeing dinosaur footsteps, mammals within whom humans were not even dreams. The tree died and got buried and ticked time not with DNA but with minerals, and one day it was resurrected by the strange limbs of an animal now fearful of the shadows of the universe, leaping not into a tree but into the wooden-cross faith that the universe is ruled by benevolence and that death could not be the end and that humans would be resurrected into a new light.

I set Nathan's blanket beside him, picked up a piece of paper, crumpled it into a ball, and placed it in his left hand. He tried to toss it, but it basically fell to the floor. I picked it up and gave it back, and he tossed it several feet. I returned it again and he tossed it farther. He smiled. On this toss the ball rolled under the huge old Santa Fe Railway safe, and I had to scoot on the floor to pull it out, which amused Scooter. By now he had scooted to the edge of the desk. I sat back in the chair and handed him the ball. Milo started to say something to him, and with his bad ears Nathan turned toward him, turned too quickly, turned right out of his limited balance, and he teetered. Toward a hard floor supporting Navajo sandpaintings showing Mother Earth and Father Sky and Black God's intended but coyote-ruined order for the universe, toward the real and hard Mother Earth, into the gravity that made standing up a screaming terror and that would seize a car racing off Hopi Point and accelerate it downward. To Milo's sudden helpless fear, Nathan teetered and started to fall off the desk.

I was the only one in a position to stop him from hitting the floor, from hitting his head and being damaged even worse, from dying right now. I had only an instant to act, and it was also an instant of deciding the value of a human life, whether his dying right now was better than twenty years of struggling and suffering, all for nothing, or whether this life was worth living. The same damned gods who had placed me in charge of rabbit and coyote lives was now demanding I determine the value of human life.

Nathan started to fall.

THE FALL

On a Grand Canyon river trip you fall about two thousand feet. This is eleven times the drop of Niagara Falls. People going over Niagara Falls in a barrel were considered fools, crazy, or suicidal, and most of them died. Two thousand feet is three times the height of Hoover Dam trying to stop the river from falling. Two thousand feet is nearly half the drop from the canyon rim to the bottom. Your river fall is spread out over more than two hundred miles, but some rapids still have the look and sound and feel of waterfalls. Hance Rapid drops thirty feet, like jumping off the roof of a three-story building, if more spread out. You are not only being pulled by gravity but pushed by gravity, for gravity has activated the strength of water.

THE ABYSS

You are falling in the same hands that guide the planets around the sun, that gave Earth the steadiness and the spin to sustain life for eons. You can feel in your cells the force that steers stars around the galaxy. Westbound, you are the moon setting in the west. In the water you can feel the falling rain and mountain cascades, become your falling. You are water sinking into the soil and meeting roots that, just like you, fully understand the laws of gravity and how to counteract gravity, how to raise water up a tree trunk and into leaves and into an apple that will fall, with seeds that perpetuate the laws of anti-gravity.

An apple falls. It is trying to reach the fertile soil but instead lands on the fertile mind of Isaac, who has a revelation. The entire solar system with all its complicated motions is organized by one simple force acting with highly reliable, mathematical precision. The planets are trying to fly off into cold space, but falling keeps them in steady orbits, keeps Earth warm and rain falling and apples growing. Isaac also sees that the orderly and benevolent solar system could come only from a supreme and orderly and benevolent mind. When you watch the moon set, when you bite into an apple and feel its juices nurturing you, you are tasting the glory of God.

Isaac is so excited by his revelation that he waves over a man and a woman walking by. He tells them that the night sky is the face of God, the setting moon his eye winking. He tells them that rivers are the blood vessels of God, and that through apples you can taste and welcome God's blessings. Isaac hands them his apple.

The couple tastes the apple, and suddenly they see they are naked, and they are ashamed. They see gravity crashing asteroids into Earth and extinguishing millions of species. They see rain becoming floods drowning innocent animals, with no Ark to save them. They see people falling into the Colorado River and being swept away and drowned. They see carpenters falling off ladders, babies falling from cribs, airplanes falling from the sky, people clutching hearts and wounds and falling dead. They see caskets falling into graves. They realize it was themselves, by biting into the apple of good and evil, who had introduced evil and suffering and death into a world that was designed to be perfect. They'd been tricked. They get very angry at Isaac, accuse him of being Satan, and threaten to cut out his heart.

No, answers Isaac, I am doing God's work, and so are you. You have just done God an enormous favor, taking the blame for an imperfect world off his shoulders and putting it upon yourselves, upon foolish humans everywhere. When you opened your mouths to bite the apple you opened the loophole through which God can escape from the trap of monotheism; now he gets the glory of being all-powerful and all-good but none of the blame for allowing evil. The theological necessity for this is just as inescapable as the mathematical laws of gravity. God is most grateful to you and will save a garden for you in heaven.

This is the universe in which we are falling, this is the river. No one blames gravity for doing its job; we rely on gravity to be reliable. No one blames gravity when a shooting star burns out. No one expects gravity to care. No one blames the rain for falling downward, for having its own reliable nature, quenching thirst, growing apples. No one expects water to care.

The fallen world is complicated by caring, or lack of caring. The fallen world is chaotic with selfishness, greed, meanness, dishonesty, lust, and violence, all of which have to be regulated by caring, but human caring thins out quickly as it moves away from the self, moves from family to friends to village to tribe, and when it steps beyond tribes it often turns into animosity. Another symptom of human selfishness is that most of the gods humans invented were tribal gods who created and ruled their own people but left other tribes to fend for themselves. Adam and Eve were a theological breakthrough, for now only one God created only one human, the progenitor not of any tribe but of all humans, and now the only one God did not have an animal shape and spirit but had human shape and gave his shape to humans, meaning that all humans contained God and that to honor God you were supposed to honor all humans. Because the tribal world had become a world of empires, a world more brutal, humans needed a larger human identity and a larger morality and a larger god. Yet humans remain biologically selfish, and universal moralities don't always reach deep, even when enforced by a righteous god.

This stream of thought does come back around to dropping two

THE ABYSS

thousand feet. Museum guides, birding guides, and fishing guides seldom face serious risks, but kayaking guides inevitably deal with dangerous situations, maybe even life-and-death situations. When someone gets into trouble, you have to decide how much risk you will take compared with how much risk someone is in, someone who was a total stranger at the start of the trip. But why should it make any difference whether someone is a stranger or a friend? Amid a chaos of falling and swirling water, amid a chaos of human selfishness and caring, you are forced to pass judgment on the value of human lives. But who has appointed me this judge, and who is deciding the value of my life?

HOPI POINT

It was always Hopi Point. Milo never mentioned any other place for committing suicide. To get to Hopi Point he would need to drive five miles and pass Pima Point and Mojave Point, named for other tribes and spiritualities. He would have to pass the Abyss. Hopi Point had more stone walls and railings and trees than many places, making it trickier to drive his car to the rim. Monica's antique car. They had separated, but she had left her car behind, a constant reminder. The Abyss offered a long stretch where the road was almost on the edge, with few obstructions. I never asked Milo why it had to be Hopi Point. We sort of understood.

Hopi Point does have a wonderful view, including a longer stretch of the Colorado River than most overlooks, and beautiful sunsets. More tellingly, Hopi Point evokes a deep sense of belonging and spirituality. It was Hopi ancestors a thousand years ago who made the canyon more of a home than it ever would be for today's residents, and made the canyon the place of creation and the place where souls journey to reach the Sipapu and the afterlife. Perhaps it was healthier that Milo was fixated on Hopi Point and not the Abyss.

The Hopis have a stronger sense of an orderly and benevolent cosmos than do many other tribes, including the Navajos. The Hopis are farmers, for whom the cosmos consists of nurturing soil and rain and seeds that unfold life in highly reliable patterns. But because the Hopis are farmers in a desert, they never take rain and harvests for granted, and they feel a deeper than usual gratitude when life does come through. Hopi gods are mostly benevolent and visit Hopi villages to join highly elaborate ceremonies to celebrate the patterns and gifts of life. The Navajos for most of their history were hunters and gatherers for whom food was far less reliable, food often running away and sometimes turning and attacking, and the Navajo universe is similarly dangerous, full of malicious spirits who need to be placated. For the Navajos, coyotes were competitors for food, and Coyote is a major and negative force. But coyotes don't seek corn, so for the Hopis Coyote is harmless, a mere clown. For the Hopis, too, Coyote helped place the stars in the sky, but this was a decorating, not a despoiling of creation.

Milo said that after he attended ceremonies at a Hopi village he heard the drumming for days afterward, even in his sleep, that powerful, rhythmic, hypnotic drumming and chanting.

The drumming was a mighty heartbeat, the heartbeat of a deer who was browsing near a spring and heard a sound that was wrong, dangerously wrong, and her heart drummed and her mind surged into alertness, and she saw every tree and boulder, every color and texture, with acute vividness, saw them with the sacred value of her own life, and suddenly she saw nothing, for an arrow had broken her heart. Now her skin had become drums and was pulsing again; human heartbeats and arms brought her back to life. From mountain slopes, from mesa shadows, from canyon rims, from desert washes, many deer were summoned into the plaza, their sacrificial blood pumping the katsina dancers onward, their large brown eyes hypnotizing the crowd, who began to feel the deer within themselves, their shared life and wildness and gratitude, who began to see with large brown eyes and sacred alertness every color and texture of rocks and sky, of the mysterious canyon out of which life had emerged and into which the dead would return. With this thunder the katsinas inspired thunderclouds to dance and grant rain to the earth; with this drumming humans could merge their tiny and fragile heartbeats and helpless wills into the powerfully drumming sources of all life.

The Hopis do leave ambiguity in their cosmos. Instead of one benevolent god and one Satan who fight each other, the Hopis have

THE ABYSS

Maasaw, who is both the god of death and the most important bene-factor of humans. Maasaw can have a ghastly, bloody face or a young, handsome face. Maasaw was the owner of the upper, fourth world and welcomed humans into it and gave them fire and crops and animals. When people die, Maasaw decides if they are worthy of an afterlife. Evildoers may face punishment, but there is no Hell. The dead play a crucial role in the cosmos, sending rainclouds to the Hopi villages. The supernatural exists to support the natural world, which may be troubled but is fundamentally orderly and nurturing. The ambiguity of the Hopi cosmos is symbolized by rattlesnakes, which are both naturally dangerous and supernaturally the bringers of rain.

Hermit's Rest once got a lesson in the ambiguity of rattlesnakes. One day a tourist rushed in screaming about a rattlesnake outside. One employee, Harry, had grown up in his parents' exotic pet shop and knew how to handle all kinds of critters. To save the public, or maybe to save an innocent snake from hysterical tourists, Harry went out and, with bare hands, gently picked up the snake, which looped freely over his arms. It was a Grand Canyon subspecies more docile than diamondbacks. Harry explained that it was a common mis-conception that the safest way to handle a snake is to grip it behind the head, for this makes it angry, and you don't want to make a rat-tlesnake angry. Harry kept the snake as a pet for a while, sometimes letting it roam around his trailer, and named it "Hopi."

There was one catch about killing yourself at Hopi Point. The side canyon beneath it is called "The Inferno." The Christian Hell. You will pass from the cosmos of one desert tribe into the cosmos of a very different desert tribe. The Hopis were one of the most isolated tribes in North America and never had to worry about Roman, Egyptian, and Babylonian armies marching through and crushing them. They mainly worried about rain and corn. The Jews, too, in their desert, may have appreciated harvests more than did Edenic peoples, but for them survival was mainly about the human world, with its massive inequities, with the strong crushing the weak, requiring a protective god stronger than the Egyptian gods who brought Nile floods. For the Jews and subsequent Christians, rain was forty days and forty nights of rain to cleanse the world of inequities, even if it drowned crops and a million gentle deer who only wanted to live. Let justice roll down like rain, and righteousness like a mighty stream. There is a river. To enforce righteousness amid such chaos, God had to be highly punitive, setting up a Hell full of torments. In Dante's version, Hell has nine levels and you have to go below most of them, even below murderers, to find the suicides. The souls of those who have killed themselves are denied human form and imprisoned in decrepit trees with black leaves and poisonous thorns. Perched in the trees are harpies with huge dark wings, feathered bellies, clawed feet, and monstrous screeches. The harpies constantly gnaw on the trees, making them bleed and scream. When Judgment Day reunites the souls and bodies of others, suicides will not be given back what they have thrown away.

For suicide to deserve worse punishment than murder, for the Christian church to support laws that the corpse of a suicide should be punished by hanging, whipping, beating, burning, or abandonment in public dumps full of vultures, meant that suicide is a crime against God. Suicide denies that creation is good. It desecrates God's own image. It ignores God's promise of help for suffering. It declares that God's universe is chaos, evil, even Hell, not worth living in. It contradicts in blood heavier than ink the tomes theologians piled up trying to justify a perfect God's creation of an imperfect world. It seized for human hands a judgment day frighteningly unsafe even in the inscrutable mind of God, who would be forced to seek to solve the paradoxes of his own identity and creation by committing the paradox and the sin of committing suicide on his cross.

THE INFERNO

I was in Hell. More specifically, I was camping in the Inferno. At Hopi Point far above me tiny lights flickered, cameras flashed as if they could compensate for the waning sunset light and illuminate the entire canyon and sky, flashed like tiny human egos in a vast universe.

Topographer Francois Matthes named the Inferno in 1906 while surveying it during a summer heat wave, with a *low* temperature of 96. In an era when people read plenty of classical literature, many tourists readily understood "Inferno" to refer to Dante's *Inferno*, and they were comparing Dante and the canyon, especially the most famous artwork of Dante's *Inferno*, Gustave Doré's canyon-like scenes, with cliffs, peaks, boulders, perilous passageways, mythic rivers, deep gloom, and the souls of the dead. Indeed, many early visitors viewed the canyon as a hellish landscape, for it defied the models of natural beauty Americans had inherited from Europe, offering not green pastures with lakes but naked rocks with bizarre shapes and colors. Many visitors found the canyon disturbing not just aesthetically but spiritually; it was a vision of an inhuman universe, the incarnation of new scientific ideas about deep and strange space, time, and evolution. A number of canyon places got names with "Hell" or "Devil" in them.

In Dante's time, Hell was thoroughly real and terrifying. Dante drew on Christian tradition and his own imagination to create a Hell elaborate with places, architecture, theology, mythology, symbolism, characters, and events. Instead of the traditional hellfire throughout, Dante invented punishments appropriate to each sin, such as a swamp of garbage and feces in which gluttons are trapped. Dante's Hell was a deep underground pit with nine circles, or tiers, tapering toward the bottom, where Satan lived. It was, indeed, a good match for the Grand Canyon, which has nine circles, or strata, tapering toward a final layer, the schist, black and twisted, cooked by ancient underground fires, and very hot in the summer. At the bottom of Dante's Hell is a river, Lethe, which erodes all memory.

Seeing the Grand Canyon as Hell would not have been an act of imagination for the Gnostics, early Christianity challengers to the theology that God and the creation are good. The Gnostics held that creation is outright evil, created in confusion by a demiurge who was ignorant of the existence of a higher God. Christ came to Earth not to redeem a good creation from Adam's errors but to descend into Hell to give humans the secret knowledge of how to find the good God.

In the Grand Canyon the Gnostics would have seen plenty of evidence for their theology. The canyon is chaos, its cliffs cracked and broken and collapsed, its boulders and rubble strewn mindlessly, leaving no easy passages or living places for humans. The canyon's

CANYON AND COSMOS

heat and dryness are demonic. The canyon's plants are often forced to grow out of rocks, and many wear crowns of thorns to torment hungry and thirsty animals. The canyon is full of serpents, reptiles, scorpions, and insects armed with poison and fangs and stingers for tormenting other creatures. The Gnostics would point out that the rock stratum on which I rested, the Bright Angel Shale, contained no angels, no help, no hope, that it was only a dark rock made out of mud, brittle and broken.

I stirred uncomfortably on the hard ground. I wondered why Dante had exempted Democritus from hellfire. Dante had placed some of the Greek philosophers and poets he admired in Limbo, including Aristotle, whom Dante admired so much he organized his Hell by Aristotle's classification of sins rather than by the Church's own system. Limbo also held Democritus, whom Dante disliked because he "ascribes the world to chance."

At a time when most Greeks accepted the Greek gods as thoroughly real, philosophers began toying with ideas that the world consisted of natural forces like earth, air, fire, and water, and Democritus proposed that matter is built of more elementary building blocks—atoms. Smooth atoms slipped past one another and formed water, while rough atoms hooked together and formed rocks. Atoms in motion combined into all the forms and forces and events of the universe. The universe had existed forever and required no gods. Plato said Democritus's books should be burned.

Drifting into dreamland, I fantasized about what would happen if Democritus and not Virgil guided Dante's descent into the Grand Canyon.

Democritus pauses at the Kaibab Limestone and points to some rock blobs within it, fossilized sponges, a matter of special interest to Greeks, the world's best sponge divers. Sponges are a good example of how atoms can sponge up atoms and build order. But Dante points out that Stone Age humans turned fossil sponges—chert—into arrowheads and knives for murdering other people. Only the Christian Hell can turn murder into justice.

Democritus points to a slab of Coconino Sandstone, with tracks where a lizard climbed a sand dune to sun itself. That little

THE ABYSS

sun-worshipping impulse, translated into Egyptian temples someday, would prompt Dante to cast pagans into Hell. But Dante rebuts that Democritus's pet atoms formed eons of mindless dinosaur bellowing and farting, redeemed by human souls and divine justice. Democritus points to the Hermit Shale with the tracks of amphibians on their way to becoming mammal philosophers and poets. Dante points at Hermit boulders scattered chaotically. In his Inferno much of the rubble was caused by the earthquake that shook all Earth when Christ died and delivered salvation to an eroded creation. What does Democritus have to offer? Nothing but erosion, massive erosion, a world falling apart.

Democritus crunches over the Bright Angel Shale, full of trilobite tracks, and says, "Let there be light." When trilobites came up with the first good eyes, they set the course to a god made of light and to angels bright and not dark, and to lightning bolts humans misinterpreted as acts of justice. The Greek kosmos requires no god to make it, no fear of lightning messages, no fear of death and punishment, and still offers Aegean sparkles and grapes. Dante replies that human society long ago ceased being tribal villages where everyone knew one another, where peer pressure and a bit of katsina pressure was enough to regulate human behaviors. Human society became a world of strangers and empires where the desires nature implanted in us turn into nightmare chaos. God is saving nature from itself. Adam is saving atoms from themselves. God allows us to praise your atoms and grapes.

They reach the black schist, where Democritus sees a great river and Dante sees rivers of tears in Hell.

That night, atop my rocky bed atop the black cliffs, I tossed and turned.

In the morning, I saw Satan. As in church murals, he had curved horns and cloven hoofs. Satan stared at me, but didn't seem worried. The local desert bighorn sheep were accustomed to people and knew how to avoid them. This bighorn stepped down some ledges, carefully but spryly, into the creekbed and up the other side. He stopped and stared at me again. He was telling me: I am thoroughly at home here. What about you?

Ruins

When we stopped at the Ancestral Puebloan ruins on the Unkar Creek delta, Dale felt at home. One of his best canyon adventures was rediscovering the route by which the Puebloans migrated from the Unkar delta to the North Rim every spring. With all its disadvantages, the canyon offered a unique combination of climates in a compact space, allowing year-round farming, the inner canyon in winter and the rim in summer. Their migration route had been forgotten, and Dale figured it out, partly from a topographical map and partly from trial-and-error hiking and climbing.

We walked around the ruins, one of the largest complexes of ruins in the canyon, occupied for some 350 years. We admired lots of potsherds of various designs, what remained of vessels once used to store seeds and food and to carry water. We were reminded that the canyon had once been someone's home, that people had spent their entire lives here and never imagined other worlds. We felt the ghosts around us. Our brief visit here, and our whole river trip, reminded us that we were only passing through.

In his rediscovery of this ancient migration route, Dale had connected his own North Rim house with these Unkar houses and connected his life with these lives. Yet despite all the Puebloans' years here, despite all their skills, forces far more powerful than themselves decided that they had to disappear.

The Sublime

There's something about October light in the canyon. Professional photographers agree that October light sets the cliffs glowing with deeper intensity and colors. It's something about the angle of the sun and the lensing of the atmosphere. It's something about beauty. It's something about humans loving being alive on this Earth.

Our October kayak trip gave us some of the deepest canyon beauty I'd ever seen. The hour before sunset was the best, bringing out the veins of speckled granite winding through the polished schist, bringing out the energy of the river, bringing out the blueness of the sky. As shadows climbed the cliffs, the cliffs became vivid candles, votive

THE ABYSS

candles. This light was "the sublime" that Romantic landscape painters hoped to portray.

Our entire journey would be bathed in October light, whether we noticed it or not. Often, we would not. We studied the October light merely for what it said about the routes and hazards in the river. Yet even when we saw nothing but difficulties and felt nothing but worry, our faces were glowing sublimely with October light that was happening long before we were born and that would continue long after we were gone, October light pouring from every star in the cosmos.

The river dropped fast during the night. When we'd tied up the rafts, we studied the water beneath them to avoid waking up and finding them draped over rocks, but the water fell more than we expected and one raft was indeed tilted on a rock. We'd camped a mile above Hance Rapid, normally the rockiest rapid in the canyon—and one of the deadliest—and now it was a gauntlet of boulders, a demoralizing beginning to days of low water. Our raft guides had a hard time believing this was really 5,000 cfs and thought it had to be lower, more like 3,000 cfs. Later, we learned that it probably was. At higher flows Hance offered several options, but today we'd be forced to start on one side and maneuver across the entire, wide rapid, which unnerved me.

Directly above us on the rim was Moran Point, named for Thomas Moran, the artist whose 1873 painting of the canyon was hung in the US Capitol, defining the canyon for millions of people. Moran lived in the universe of Romanticism, in which nature was not just pretty scenery or geological stories or national icons but the subtle face of divinity. When Newton evicted God from comets and Luther evicted God from statues and Voltaire evicted God from miracles and even theologians began evicting God from the Bible, God took refuge in Alpine peaks and chasms, in rainbows and mist and waterfalls, in nature's power and majesty, inviting you to meet him not rationally but through feelings, through the sublime.

Thomas Moran wanted to turn the canyon into the sublime, but he had to fake it, ignoring its geological strata and turning it into a landscape of peaks, just like the Alps; ignoring its true, odd colors and giving the whole canyon a golden Italian hue; adding rainbows and sunbeams and mist that were standard symbols of the mystical. Moran showed no signs of disharmony, no mountain lions attacking deer. If Moran showed the river at all, it was only a pretty silver thread. Many of the tourists at Moran Point were seeing Moran's divinity. Even many people who had discarded God—but kept his prophets Thoreau and Muir—had perpetuated Romanticism's faith that nature is beautiful, peaceful, harmonious, and wise.

We were not so sure. We were not seeing any pretty silver thread. We were studying the aftermath of deluges, often without rainbows, that had swept monsters into the river, boulders the river attacked angrily. We were feeling small and fragile and fearful. We were struggling through the sublime.

I have thought of a kayak paddle as a paintbrush, with its long wooden shaft flaring at its end. I dip it into color. I paint beautiful ripples. The cliffs reflected in the water also ripple. Or are these worry lines? Looking at Hance Rapid, I cannot afford to imagine harmony there, only dangerous obstacles, a brush with death. If you are hiking canyon trails in summer and you begin to see the sun glowing with divinity, you are likely suffering from sunstroke. The canyon forces you out of the world of charming ideas and into its own hard realities. The waves in Hance are really Edvard Munch waves, and if you flip you'd better restrain any impulse to scream.

We did make it. Not harmoniously.

The Underworld

Where should I begin looking for Ron "in the river"? His only known location was at Grapevine Rapid at river mile 82. But when we got to the confluence with the Little Colorado River, twenty miles upstream from Grapevine, I wondered if I should begin looking. Up the Little Colorado was the Sipapu, where Hopi souls travel to reach the afterlife. I had met Ron at a ranger talk about the canyon's spiritual meanings for Hopis. He seemed to admire Hopi spirituality. I thought of how Milo was ready to travel miles out of his way to kill himself at Hopi Point. Ron would need to make more of an effort to get to the Sipapu, but it might give his death some poetry. Maybe I should at least start training my eyes to notice unusual shapes in the river. It wasn't unusual to see dead fish floating along, but I'd hardly ever seen larger animals. One time I passed a dead bighorn sheep, which I guessed had stepped on a loose rock and fallen. But now that the abstract idea of looking for Ron was becoming the real possibility of finding him, it made me uneasy. I recalled someone telling me how he was kayaking on a mountain lake and saw a dead, bloated deer floating there, and from curiosity he approached it and gently tapped it with the tip of his kayak and the deer exploded, throwing pus and flesh and stench all over him.

CRASH

At the Little Colorado River confluence, not far from the Sipapu entrance to the underworld, there loomed above us a butte smeared with carbon atoms that only a few decades before had been human, or possibly part of a diamond ring.

Some 350 million years ago many of these carbon atoms were flowing through the ocean as fish. Fish felt a moment of dread and pain and became shark flesh and shark motion and shark feelings, betraying their former selves by attacking fish, even their own fish children, being loyal to the game of life by showing no loyalty to the players. Some carbon atoms drifted to the seabed and were buried by hundreds of feet of carbon and turned into rock, which eventually rose high into cliffs, from which some carbon atoms escaped and rejoined life, life's new shapes playing the same old game.

One day, some of those escaped carbon atoms, now human, returned to the Grand Canyon cliffs and the ancient ocean that had long been their home; so shouldn't they have seen these cliffs as a homecoming?

Rim tourists view these cliffs as a thing of beauty but don't glimpse the tragedies hidden there, how tsunamis had swept through this ocean and buried fish and sharks, which struggled for a moment and died, leaving a long cemetery of molecules dialed to terror. Every day, this ocean held little accidents and deaths. Life spent 350 million years improving its eyes and limbs and reactions to avoid accidents, yet it

CANYON AND COSMOS

had also continued coming up with new possibilities for accidents. These returning carbon atoms, now human, expected to see the cliffs as a thing of beauty. But against vast odds, in a vast and almost empty sky, in a sky also apparently empty of logic and God, two airplanes full of highly evolved eyes did not see each other and crossed paths at more than three hundred miles per hour, the wing of one plane cutting off the tail of the other plane. Stripped of eons of aerodynamics and relapsed to elementary physics, to the gravity that crashed asteroids onto Earth and gouged out craters, the planes fell. People were stripped of thousands of years of words for expressing every event and feeling and found they were animals with a simple, ancient mechanism for expressing terror: screaming. As the cliffs gyrated and raced closer, people screamed from envisioning the plane slamming and shattering, their bodies being crushed and torn, gasoline gushing out and exploding, fire raging, bodies burning, faces melting, the hair and clothing they had chosen so carefully to define themselves burning away, the genitals and breasts that connected them to one another and to the future melting away, their powerful bonds to the living cut forever, their water boiling skyward but not quite like souls, their charred bodies disgusting even the vultures. Against all of this, they screamed.

Perhaps their screaming was a true vision of the canyon, of cliffs made of death, endless deaths over tens of millions of years, coral deaths quiet and oblivious, fish deaths with tiny bursts of terror and pain, building cliffs solid with terror and pain, terror and pain felt over and over again by carbon atoms living and dying over and over again, terror and pain these now-human atoms had forgotten and left behind but seemed to suddenly remember now as if in that flashback of your whole life when you are about to die, this a flashback through vast reincarnations, as if the terrors buried within these cliffs had suddenly found mouths through which to be released and to announce their true verdict on creation as a thing of horror.

Or perhaps their screaming was a final affirmation of creation, a judgment carefully learned and strengthened through eons of evolution, the value of being biological forms and identities, of having flesh that feels the earth and the breeze, of having eyes that see the colors

182

of the trees and the patterns of the stars, of having maze minds to run the world through, of having unique lives never lived before and never again, a value they worked hard to maintain and enhance and which they would not surrender willingly or sooner than required. Their screaming was horror at having to lose their forms and abilities and individualities and to melt back into chaos and anonymity, back into the primitive ocean they thought they had escaped, horror that the same sharks' teeth that had killed them long ago were about to taste their blood again. Their screaming was a final offering, an older version of the chants that filled temples and cathedrals with praise for being here and that had turned a nearby canyon spring into the Sipapu from which life had emerged and through which life would not have to die.

This butte was across from Desert View, where George Ritchey had envisioned the world's greatest observatory. In 1928 he drew up an illustration of his observatory, and in the background was the butte on which the plane would crash decades later. Ritchey's biographer, Donald Osterbrock, used this illustration on his book cover. Donald's mother, Elsie, was a passenger on the plane that hit the butte. Donald's son, as a boy, suffered a severe bicycle accident and head injury that left him permanently limited and living with his parents. Donald, the director of Lick Observatory, did pioneering work on the violence at the heart of galaxies. One day walking to work, Donald suffered heart violence and died, without saying goodbye to the universe. What would the Desert View observatory, meant to spy into the deepest nature of the universe, have reported about all this? Its astronomers could have published a graph plotting the jagged and spotty relationship between order and absurdity.

The dead from one plane, sixty-six bodies, almost all unidentified, were buried together in the Flagstaff, Arizona, cemetery. Thirteen years later, in a family plot right next to them, was buried Vesto Slipher, an astronomer at Flagstaff's Lowell Observatory. Slipher's spectrographic studies of spiral galaxies revealed extreme motions, the motions Edwin Hubble soon used to prove that the universe is expanding. Somehow, the motions of galaxies had turned into the motions of cells and bodies, then into the motions of machines, then into the motions of machines colliding and crashing and igniting bodies. From his grave Vesto Slipher, now motionless, could study the zero redshifts of sixty-six bodies, now slowly contracting, and seek what they implied about the universe.

WHERE IS THY STING?

Across the river from us the most prominent landmark was Buddha Temple. Most rim tourists don't realize this butte is named Buddha Temple, and Buddha might prefer it this way, prefer that his butte be a powerful reality not masked by human concepts. For Buddha, the canyon is full of glorious shapes and colors only because it is a glorious emptiness, created by a never-striving river. Having taught compassion for all life, Buddha might appreciate that Buddha Temple is full of life, its Redwall Limestone base full of fossils, from the era when insects were getting started.

I was standing alongside the Bright Angel Trail with Larry, the canyon's leading biologist. He pointed out an orange shape flying by, a wasp called a tarantula hawk. He told of how, when he was new to the canyon and knew a lot less about tarantula hawks, he had pulled ashore to scout Lava Falls, only his second time there. A friend saw a tarantula hawk flying past and snagged it with his hat and, assuming Larry would be interested, handed the hat to him. Larry, thoroughly preoccupied by the rapid, held the hat and wasp in his hand and was fascinated to see a stinger emerging from his flesh, penetrating right through his hand. For a moment he retained enough rationality to think: "This is really going to hurt." It really did. A quiet little bug had obliterated all thoughts of Lava Falls.

With Buddha Temple silent and listening, Larry explained that the tarantula hawk is one of the largest of wasps, with a long stinger that delivers one of the most powerful and painful stings in the insect world. The wasp feeds on flowers and uses its stinger mainly to launch its offspring. The female wasp stings a tarantula, paralyzing it. She grasps the tarantula with hooked-claw legs and drags it to a carefully dug hole, lays one egg on the tarantula's abdomen, and buries it. The egg becomes a larva that burrows into the still-living, still-paralyzed spider and starts eating it. The larva seems to be programmed to avoid eating the spider's vital organs, to keep the tarantula alive as long as possible. The tarantula contorts as if feeling the pain of being devoured alive, and finally dies. The new wasp emerges into the pretty flower world. The tarantula hawk's strategy was so productive that it has hundreds of species, called ichneumonids. Some rely on caterpillars or crickets.

People who have seen too many Hollywood movies may be disinclined to sympathize with tarantulas, for Hollywood has cast them as monsters creeping across sleeping human bodies. But in truth tarantulas never go hunting for large animals, and tarantula poison is relatively mild, incapable of killing a human, hurting only mildly. I once tested this theory when I saw a tarantula strolling along a busy Grand Canyon rim sidewalk, at high risk of being stepped on, perhaps deliberately by little boys. I tried to herd the tarantula off the sidewalk with my foot, and when this failed I poked it with my finger. The tarantula did not seem to appreciate my act of transspecies compassion and only continued along the sidewalk. Soon I was almost petting it. Did Buddha recognize my act of compassion?

In the decades after Darwin poked God off his golden sidewalk and into the dark woods, vampire wasps became a major case study in the debate over the goodness or evil of the natural world. Vampire wasps could be seen as proof of not only cosmic indifference but of outright malice, of existential horror. Leading Christian thinkers were determined to avoid this and to turn vampire wasps into evidence of God's planning and benevolence. Three decades before Darwin, Charles Lyell recognized that his *Principles of Geology* was encouraging an atheist universe, and he tried to salvage God's role by arguing that vampire wasps were part of God's plan to limit insect populations for the benefit of human agriculture. But when Darwin, in his *On the Origin of Species*, wanted to prove that natural selection is a thoroughly selfish process, he brought up vampire wasps. After Darwin, William Kirby, a prominent British entomologist and churchman, argued that vampire wasps proved that a careful and generous design was behind even wasp motherhood. Naturalist Asa Gray wrote to Darwin that even if natural selection was true, there had to be a moral purpose to its results, but Darwin replied that he could not believe that a benevolent God would have created vampire wasps and similar cruelties. Mark Twain wrote a short story, "Little Bessie Would Assist Providence," in which a mother is arguing that God must have had a good reason for giving typhus to Bessie's friend Billy, and Bessie answers sarcastically that tarantula hawks must be devouring spiders to make them devout, make them "praise God for His infinite mercies," upon which Bessie's mother faints.

One twilight, camping at the canyon bottom, I noticed a dense spider web on a bush. A small moth had flown into it and was fluttering on it, struggling, baffled by a new reality. I sympathized with the moth, its unhappiness, its struggle, its love of flying through the canyon twilight, its desire to live. The moth was just myself in a smaller package. It would be easy for me to help the moth escape, and the greatest gift for the moth, even if I was merely another inexplicable force the moth would not thank. I picked up a twig, a scalpel for the surgery. Where was the spider? Surely the moth's many radar-screen pings alerted the spider.

The spider. I hadn't even thought about the spider. Or maybe I was unthinking about the spider with the mammalian programming that profiled spiders and snakes as dangerous and led Hollywood to turn tarantulas into aggressive monsters. Unseen, the spider had remained an idea and not an individual life. Hadn't I once reached out to a tarantula to save its life? Wasn't this spider simply trying to go on living? It hadn't chosen to be a spider but was conscripted into a form that had no other skills for surviving. It had taken a lot of trouble to weave this web, an amazing and pretty mandala, which I was now planning to ruin. Who had given me the right? The weaving skills of spiders had impressed humans into promoting spiders into creative deities, even world creators. For the Hopis and probably for their Grand Canyon ancestors who watched the ancestors of this spider, Spider Woman or Spider Grandmother had guided humans out of the darkness and wrongness of the lower worlds and into the light and fertility of this world. For the Navajos, Spider Woman empowered the hero twins to rid the world of evil and taught the Navajos how to weave amazing rugs. So who was I to punish Spider Woman? It wasn't my fault that a dumb moth had flown into her web.

THE ABYSS

But was it the moth's fault that it hadn't been ennobled in human mythology? What if this was not a moth but a butterfly, whose biological biography was not much different from a moth's but who got far more respect from humans? Poets, artists, and songwriters selected butterflies and not moths as images of beauty and happiness. Prophets chose butterflies as symbols of souls and resurrection. Butterflies are more colorful and humans are easily amused by colors, feeling far more valuable if they adorn themselves with gold and jewels. Was I going to allow tawdry human tastes to decide a creature's fate?

The moth continued struggling but only entangled one of its wings worse. Why was I hesitating to release it from its misery? Why allow compassion to be arrested by intellectual turbidity? I could be a moth god performing a miracle, creating dozens of generations of moths. Then again, I could force a spider to go hungry for days. I could postpone or diminish or entirely cancel a next generation of spiders. I couldn't decide what to do. Maybe I should only watch nature at work, doing what it does a million times every second, setting up life-and-death struggles. Creatures had no choice but to kill and devour one another. All this misery was the fault of the second law of thermodynamics, which requires energy to dissipate and forces animals to constantly find new energy, even if they have to steal it from one another. Why couldn't the laws of nature have been different, more generous? One law of physics, unjustifiable even as physics, had generated a monstrous universe. In this context, in this chaos, in all this misery, my actions wouldn't make any difference, and my ability to choose might be a luxury and my compassion meaningless. Buddha Temple, with its stone heart, might prescribe compassion, but it assumed a karma that justified the shapes and sufferings and destinies of spiders and moths and people. Other religions, too, with Kronos and Kali, monsters and demons, deluges and Ragnarok, acknowledged that the entire universe devours itself. The universe couldn't decide what to do.

The night arrived. I couldn't see if the spider had arrived or if the moth was still struggling. Above me, as I slept, spanned the Milky Way, which for the Akimel O'Odham and Maricopa tribes of southern

187

Arizona is a spider web, amid which sits the spider (the constellation Cassiopeia) who had spun other webs to create the earth. All around me stars wove atoms together, strands of stars added up to our galaxy, and strands of galaxies added up to vast webs of order and light—but also vast regions of emptiness and darkness. All night long the cosmic spider wove my cells and dreams. All night long the cosmic spider wove uncountable lives and hungers and deaths, wove them out of the same silk that had formed Buddha's tears and that tied Christ to the northern cross so that his blood could balm a universe full of suffering.

When I awoke, the spider web was shining with dawn and a desert touch of dew. The moth was dead. The spider was still invisible. I still didn't know what to do.

ARISTOTLE ARGUES WITH NATHAN

We had kayaked into the mythological world. For the next forty miles most of the buttes had mythological names. We were starting in ancient Greece and Rome: Apollo Temple, Venus Temple, Jupiter Temple, Juno Temple.

Rome was founded on infanticide. The unwanted babies Romulus and Remus were thrown into the Tiber River to drown, but a she-wolf rescued and nursed them and they became the founders of Rome. In spite of this advisory that unwanted children might have value, Roman law required that all deformed children be put to death. In Athens, Aristotle, whom the future would honor as a founder of rationality and science, advocated a law requiring that all deformed Greek babies be killed. The Greeks did routinely discard deformed and unhealthy babies, tossing them onto garbage dumps or manure piles, letting them die of hunger, cold, or animal attacks. In the shadow of the Library of Alexandria, holding all of human wisdom, Egyptians watched Greeks tossing babies onto manure piles, scorned the Greeks as barbarians, and retrieved some babies—to raise as slaves.

Aristotle would have a final solution for the problem of Nathan. Aristotle would carry Nathan into the woods and dump him and let the coyotes devour him. Or if Aristotle was feeling humane he would place Nathan in a big jar to keep the animals away but not

THE ABYSS

the cold. But with a cliff so close, Aristotle might carry Nathan to the edge and toss him into the canyon. If Aristotle was sitting there when Nathan started to fall off a desk, the moral action would be to do nothing, to watch him fall and hope he cracked his skull and died. Who was I to argue with Aristotle?

In the ancient world infanticide was widespread, sanctioned by gods, customs, and laws. Deformed babies could result from evil spirits or evil deeds and still contain evil. Often, people sacrificed babies to the gods; in the Roman world, to Jupiter and Juno. But the gods were not fooled and saw that people only wanted to get rid of their babies, especially girl babies.

In the ancient world almost the only people who banned infanticide were the Jews. For them all humans were created by God, in God's image, for God's good purposes. Surely God even had a purpose in creating deformed babies, if only to refine his people's talents for mercy. If God's purposes seemed inscrutable or even deranged, you simply had to trust in God's wisdom. From Judaism, disapproval of infanticide passed into Christianity and Islam, and when the Christian nations colonized the world, they spread their prohibition against infanticide.

As the scientific worldview colonized the world, as Athens seemed a better role model than Jerusalem, what became of deformed babies? What became of all God-sanctioned mercies? Many of the moral values of the Western world, including of humanists, remain Judeo-Christian values, even if secularized into laws, and now their foundations have been removed and they are floating in the air, ready to fall. Rationalists tell us that reason alone can define the moral values that make people happy and make societies harmonious and merciful, tell us that religion was only dressing this logic in fancier and unnecessary clothes. But doesn't this logic already assume that harmony and mercy are good? Anthropologists tell us that for most human societies for most of human history, reason said that infanticide was necessary to keep populations aligned with resources, to allow parents the time, energy, and food for child-rearing. Zoologists tell us that infanticide is widespread in the animal world, including among primates, for whom child-rearing is more demanding than

189

for other animals. In humans, evolution blundered into a trap: prolonged child-rearing promoted full-time sexuality to bond parents, but full-time sexuality often produced more children than parents could handle, and produced more demands than a land and society could handle, encouraging societies to war against their neighbors. Civilization eased burdens with better harvests, better medicine, and greater affluence, but even as science technologies supported mercy, what were science ideas saying about mercy? Both Hitler and Stalin were convinced they were acting in the name of science and human progress, even if they could not agree whether science mandated hierarchy or equality; but they did agree that science mandated slaughtering millions without mercy. Did Nathan the Auschwitz SS officer result not just from the logic of reincarnation but from the logic of biology? Would Nathan the SS officer let himself, the boy, fall to the floor and die?

But wait, haven't we skipped a step here? We haven't asked all those deformed, abandoned children their philosophy about this. Listen to them crying and crying, crying out for life and mercy.

FROM THE JOURNAL OF TIKI

> I've been through caring so little about life that I wanted to die of my disease.
> I felt it was an escape, a way to get out of man's world and back to nature.
> I hated man, I thought he was a mutant.
> He destroys so much, like forests and animals, and then turns around to save lives, mine.
> It just didn't make sense, so I asked why live in such a world.

GRAPEVINE RAPID

As we approached Grapevine Rapid I paddled toward the beach above it. Maybe I was looking for a ghost. This was Ron's last known location, where the kayaker had approached him and he hadn't answered. If I was serious about finding his body, it was time for me to start paying more serious attention.

THE ABYSS

But the river was now demanding serious attention for our own sake. I asked one of our raft guides if Grapevine held anything except waves, if we should stop and scout, and he said that the only obstacle would be a hole at bottom right. This was the last time I trusted anything our raft guides said.

After getting through the big entry waves I tried to spot the hole at bottom right, but amid so much commotion there, nothing stood out. I moved leftward. I turned sideways, partly to brace through the waves but also to look upstream at the other kayakers. Here in this mythological stratum I should not have forgotten Orpheus, whose song was so beautiful that rivers changed course for it and Charon allowed him safe passage through the underworld; but Orpheus broke the rules and looked backward, and then asked for and received death. I looked forward just in time to see a sharp line right in front of me, definitely no wave, and I barely had time to straighten my boat before plunging into the hole, which backflipped me. When I rolled up I was still in the hole, side-surfing it, fitting perfectly into the trough, water cascading on one side and boiling up on the other. I was bouncing wildly but also firmly stuck.

High above Grapevine Rapid loomed Wotans Throne and Krishna Shrine on one side of the river, and Newton Butte and Lyell Butte on the other side. Wotan and Krishna were two of the world's most powerful gods, easily rescuing humans who got into any sort of trouble. Newton Butte was named in honor of Isaac Newton, who showed that planets and rivers flowed not from gods but from gravity, and Lyell Butte was named in honor of Charles Lyell, who showed that if raindrops flowed for a very long time they could build rocks and carve canyons. I had come to the canyon to explore the universe of Newton and Lyell, and now I was getting a damned close look at it, at water falling according to Newton's laws, at a boulder—probably from Lyell Butte—rolled into the river and eroding according to Lyell's laws. I struggled to keep my tilted balance, to keep the downrushing water from grabbing the edge of my boat and to keep my paddle bracing on the unsteady backwash, to stay stable when I didn't want to stay here. How did I feel about the natural world now? What good did its beauty do me now? How did the amazing balances of atoms and solar systems drop me into this? Maybe I should pray to Saint Isaac and he'd turn off gravity for a moment. Didn't I really wish I could call upon Wotan and Krishna for help? Didn't I see why they are so popular?

Before long the hole flipped me, and after some turbulence, and through no skill of my own except resignation, I got kicked out. Wotan and Krishna smirked.

ZOROASTER

Kayakers, in our more vulnerable boats, learn to pay closer attention to all the signals the river is giving us, the warning signs that can be subtle. A smooth bulge could be just a wave, but if its shape and color are slightly odd, it's more likely a boulder with only inches of water flowing over it. This attention is called "reading the river," but if the river is a book it's a book by Homer, whose father was said to be a river and who sent Odysseus on primordial, inscrutable waters. My reading skills might now prove helpful in spotting a body in the river, perhaps pinned for weeks and finally released by a weaker current.

But the signals were confusing. Many rocks had the size and rounded shapes of human heads, or skulls. Underwater algae was waving like hair. Branches bobbing in the current could be arms. I could see the river bottom, even thirty feet down, with far more shapes, blurry shapes. Until now the river had been simply a river, but now I saw it as a hiding place, a tomb.

Human bodies in this river sometimes resulted from suicides. The first known suicide drowning was in 1933, a middle-aged lady who left a note: "If you are looking for me, don't look further. I am heading for the river and am never coming back." Her body was found some three hundred miles downstream. The river gives all bodies the same funeral procession, never passing judgment on how anyone died, never pretending that life isn't really over and that decay can be defeated.

After Grapevine the next rapid was Zoroaster, full of debris from Zoroaster Temple, named for the prophet of the world's first monotheistic religion, which inspired the Jews when they encountered it in Babylon. The prophet Zoroaster decided that nature gods and tribal gods weren't good enough and that one wholly good god,

Ahura Mazda, created humans as good beings. All the world's evils came from one wholly evil god, Ahriman, who waged endless battle with Ahura Mazda. At their death, Ahura Mazda weighed humans by whether they had served good or evil. Suicide was a surrender to evil. To avoid corrupting the earth and water, Zoroastrians place dead bodies atop towers where vultures consume them. Zoroaster would not approve of dead bodies in a river.

Dead bodies floating through Zoroaster Rapid are weighed entirely by the pound, and directed not to heaven or hell but only downstream.

But wait, the name Zoroaster also gets translated as Zarathustra, which opens the possibility this rapid was named for Nietzsche's Zarathustra, actually his theft of Zarathustra for use as the prophet of no god but of human loneliness and striving and Promethean pomposity. How would this Zarathustra weigh dead bodies floating through it? Exactly the same.

Both Zarathustras noticed that life's impulse to build order and recognize patterns—and which filled the universe with gods and prophets—was at work in me now, seeking a body amid chaos, seeking an explanation Zarathustra could speak.

HORN CREEK

People do get lost in the Grand Canyon. Some people get lost and lose their lives. In 1959 a Catholic priest led two teenage boys down the Tanner Trail in the worst summer heat, carrying little water, and made a wrong turn that led to a dead end and the heatstroke death of one boy. Other hikers, out of water and imagining they can find a shortcut to the river, have dropped over ledges and ended up stranded and dead on impassible cliffs. The lost can be hard to find, as the search for Ron demonstrated.

Dale's wife Katy, also a North Rim interpretive ranger, was in charge of a dog team in that search, directing the dog trainer on and off trails. The dogs kept following the scents of deer or coyotes, and the trainer had to interpret their actions and call them back. The dogs were trained to pick up the scent of corpses in the breeze, and one time they picked up a corpse scent and followed it—to a dead deer. When one search grid was done, Katy radioed for the big military

CANYON AND COSMOS

helicopter to shuttle the dogs to the next section. The dogs seemed to enjoy the ride. For them the whole search was just a fun game in an exotic place. When I asked Katy if she knew why the search for Ron was so extensive, she said she'd found it all very odd. When I explained the vice-presidential connection, she said this had been kept secret from her and other searchers.

Dale got lost on our river trip, and never fully found himself again. If I followed Dale's logic about it, depression logic I repeatedly told him was illogical, you might say our trip contributed to his death.

Dale's rafting went pretty well, even in Hance Rapid, with only one oar. Our run in Horn Creek Rapid started out perfectly.

Horn Creek Rapid holds one of the most beautiful and astonishing waves most river runners have ever seen, anywhere. As water flows over a boulder at the top of the rapid, it crests and spreads out and curls down, glassy smooth. As the river level decreases and water is barely skimming over the boulder's bumps and grooves, the wave pulsates, dancing up and down, firing off jets that wiggle into weird shapes, flying script, as if the river is trying to write you a message, a message of wonder. It's also a message of fear. Horn Creek is the steepest drop in the canyon, and as the river level decreases, Horn Creek becomes one of the most dreaded rapids, plunging into huge holes and a cliff. Lower water also chokes off the best entrance. To avoid the holes you have to enter toward the right and move leftward, across the current, very quickly at the top, with perfect angle and timing. But the current is deceptive, and numerous boaters misjudge it and get swept rightward and into the holes and cliff.

In 1929 the park's first interpretive ranger, Glen Sturdevant, drowned in Horn Creek Rapid.

Dale and I faced Horn Creek at a moderately low level, with holes worth avoiding and one very big, steep wave unavoidable. Our position and angle and timing were perfect: I was amazed at how fast our raft tracked across the current. We plunged and faced the big wave, our angle correct. We would have made it, if only the wave hadn't broken at the worst moment. The wave cranked us upward, stopped us, made the raft tremble, punched us with a ton of water, and knocked us backward. The raft crashed down on top

THE ABYSS

of us. Violence. Confusion. I touched the shapes of an upside-down raft suddenly hostile, trying to trap me—my sandal got hooked on something and its strap got ripped loose and the sandal came off. We found the surface, climbed aboard the upside-down raft, pulled out the emergency paddle, and hobbled to shore.

Dale had never flipped a raft and was shocked.

Later, we analyzed our mistakes. We hadn't fully anticipated the cataraft's behavior. Our cargo weight was centered too far back, and I worsened it by riding in back when I should have been in front. Dale could have taken one final vigorous stroke. Otherwise, it was perfect.

I told Dale that some of the best professional river guides had flipped in Horn Creek and that guides who had flipped in less notorious rapids wished they could tell their friends they had flipped in Horn Creek. But this reasoning didn't reach Dale. He reacted as if his beloved canyon had betrayed him. Or maybe he had failed it. Or failed himself. For the rest of the trip Dale was troubled, doubting himself, asking me to confirm whitewater signs that should have been obvious to him, taking the safest runs, worrying about the unavoidables to come.

And some people find themselves. Katy was determined to run the entire river in an inflatable kayak. Before and during the trip, Dale expressed skepticism: Katy was unready, headstrong. I encouraged her to go for it. Dale deferred to my judgment, grudgingly.

Katy emerged from the trip triumphant, Dale humbled. Later, I wondered if I had helped widen the impasse already between them. Katy, too, was a highly capable person, finishing many Boston marathons, but she found few outlets for her energy in the small world of the North Rim. One year, when the North Rim closed for the winter, they stayed and watched the snow piling up. Katy's long-desired medical career wasn't going to happen here. And Dale had decided, out of some childhood hurt, that he would never have children, but Katy wanted a family. She moved on.

Dale attached excessive importance to our river trip: if only he had emerged a bold hero, his wife would not have left him. I pointed out the illogic of connecting a flipped raft in Horn Creek with your wife following her longtime dreams. Yet maybe I had a stake in denying his logic: if only I'd been riding in front, we wouldn't have flipped and Dale would have emerged a hero. Later, Dale's logic had deeper consequences: if only I'd been riding in front, Dale wouldn't have killed himself.

Horn Creek combines all the canyon's beauty and horror.

FEAR

It's why the blue heron stands so still.

She knows that the difference between life and death is so small, just one little motion. She is pretending she is dead, just another bush, so she can go on living. Living requires her to pay keen attention. She knows the eddies of the Colorado River better than anyone, all the swirls and algae and rocks that hide fish. She is waiting for the one motion that means life, to siphon that motion into her own. She has to pretend she is dead because life is dangerous and triggers fear. She has to stab quickly or the fish will dart away. She understands how the fish feel, for she feels the same way when a bobcat sneaks up on her.

She hears a splash, a steady splashing. One of those kayak things comes around the bend. In her entire life no human has tried to harm her, but she tenses and launches and flies downstream a hundred feet. The human approaches again, and she flies downstream again. Fear tells her to fly away from an approaching predator, when if she flew upstream this problem would be over.

I recognized the heron's fear within myself, for I was approaching Horn Creek Rapid at low water. As I watched the heron fleeing, I was struck by how a life-form so different from me could feel the same fear. Yet fear is as old as animal life, felt by animals I could never imagine.

Even ants know fear. On a previous kayak trip we had stopped at the beach at Phantom Ranch, two miles upstream from Horn Creek Rapid, and made lunch. Frequent eating places like this have a regular supply of crumbs and a regular supply of red ants. I must have pinched an ant between my toes: I felt a bite. It became a sharp pain, sharper and sharper. I walked to the river and stuck my foot in the cold water, which numbed the pain, but then we had to go and my

THE ABYSS

foot was back in my kayak and the pain grew steadily for two miles, grew to dominate my mind, just when I needed to concentrate on Horn Creek. I was going to get pulverized, pulverized by a stupid ant. My own nervous system, which was supposed to alert me to danger, was now placing me in danger.

At the top of the Horn Creek drainage, tunneled into the cliffs and up to the rim, was the Lost Orphan Mine, the 1950s uranium mine that Milo wanted to blame for poisoning his baby. The rim land holding the mine shaft was indeed contaminated and fenced off; the tunnels were bleeding radioactivity into the canyon; and the National Park Service warned hikers not to drink from the creek. A tiny portion of the water in Horn Creek Rapid, water that stirred such fear in boaters, came out of the creek and contained a greater fear, the fear of nations against one another, the fear that builds castles and armies and now atomic bombs, enough to destroy human civilization and eons of evolution, delivering the verdict that the whole of evolution was, in the end, insane.

On one of my backpacking trips I paused in the Horn Creek drainage and pointed out the steel mineshaft tower on the rim far above, pointed it out for someone who had been the pilot of a nuclear-armed B-52, waiting for twenty-five years to take off. I asked him if he ever had any qualms about his mission, and he said no, he was just doing his job. When I suggested that the Grand Canyon could cast things in a larger perspective, in which atomic war could ruin eons of life, he said he was just defending national security. We were surrounded by plants and insects with eons of clever strategies for conquering or exploiting or poisoning one another. I did one river trip with someone whose mathematical brilliance got him a job calculating the deaths from nuclear bombs of various megatonnage, 10,000 deaths here, 100,000 deaths there. But he had a change of heart and became a professional river guide, using his mathematical intuition to calculate the comparative rates of descent of the water and his boat into Horn Creek Rapid.

Today Horn Creek Rapid was lower than our raft guides had ever seen it. Even for kayakers the right-to-left dash was forbidding. I started to tell the kayakers it was perfectly honorable to portage a rapid, but Sandra cut me off indignantly, saying "I didn't come all this way just to walk around rapids."

We thought we saw a way to sneak past the worst of it, but once again Horn Creek's powerful currents were tricksters. Even Roy, going first, got swept off course. I didn't come close. No one else got through intact, but no one was hurt.

We ended our stressful day camping at the top of Granite Rapid, one of the most powerful and chaotic rapids, with waves slapping the cliffs so hard that people could hear it from the rim a mile above. The lower water had reduced Granite's power, but we were shocked to see a dangerous hole at the bottom. We walked down the shore and studied the hole and judged that the main current wasn't going into it, but judging the push of the current from the opposite shore of a wide rapid wasn't foolproof, and Horn Creek had just fooled us.

We were breaking a rule that some river runners take seriously: don't camp atop a serious rapid whose roar could invade your sleep and poison your dreams. Usually I enjoy listening to the river at night; it soothes me to sleep. But not tonight. Tonight the roar of Granite filled me with dread and kept me awake. I was still awake at midnight, watching the stars.

Usually I associate the stars with the soothing sounds of the river, as if I am hearing the music of the spheres, which in truth I am, for the same gravity that makes the planets roll smoothly but silently onward is also making the river flow. In the river, gravity has found a voice, Venus is singing to me, the whole night sky is whispering that it's a realm of peace. Yet tonight the music of the spheres was dissonant, violent, ominous. It was the roar of nuclear fusion across the universe. It was the roar of stars immolating themselves and destroying planets full of life. It was the forging of uranium that would roar again over Hiroshima and leave only wall shadows to weigh the value humans found in their lives. The rock debris making this noise had been swept down Monument Creek in roaring flash floods, swept down from the Abyss, for this side canyon is the Abyss. Amid the rocks the Abyss disgorged into the river were a few atoms of humans who had committed suicide in the Abyss. In Granite Rapid the Abyss had found its voice, an angry, obstructing, denying, threatening voice.

THE ABYSS

Was it from such a roar that humans have to hear their value? Past midnight, the Abyss raged against me.

At some point, I realized I had entirely forgotten about looking for Ron's body.

I did recall how the last time I'd camped here I'd hiked up Monument Creek in the evening light. I walked up a creekbed full of color, stones from every stratum above, stones rounded and polished and pretty. I walked through Vishnu Schist marbled with pink granite, the same bold granite that gave Granite Rapid its name. I entered a modest slot canyon, polished into ramps and curves and pools. From the pools, frogs sang and echoed. I reached sandstone from which little salt flowers were crystalizing, and I tasted them and found them good. I came to a terrace where Ancestral Puebloans had lived, leaving potsherds, projectile points, traces of walls, roasting pits, and smoke stains under an overhang. Monument Creek's reliable water, amid miles of drainages usually dry, probably left the Puebloans fairly content here, thanking their gods for the gifts of life, even here in the Abyss. On backpacking trips I had camped near the ruins many times. I had watched sunsets turning the Abyss cliffs gorgeous. I had hiked to the cliffs and the spring at the head of the Abyss, and one time I met a bighorn sheep, standing on a ledge, watching me, seeing our differences but also that we were the same elementary force.

The Abyss was full of ghosts, which might be hard to see in the rational daylight but seemed easier to hear at midnight.

The ghosts of the suicides, who had decided that life wasn't worth its trouble, thought they saw an opening for persuading me. Look, they said, look at all the trouble and stress you've put yourself through just today. What's the point? You didn't need to do any of this. You could be sound asleep in your own cozy bed right now, not worrying and tossing on lumpy and chilly sand. Are you doing all this just to be able to say you could have been killed today but you weren't, not today, death will have to wait a bit longer? Maybe risking your lives only confirms our verdict that life has little value. Maybe struggling down a river for two weeks is only a shorter version of a deformed child suffering for two decades only to die, of anyone struggling for a few decades for little pleasures that are not worth all the trouble.

CANYON AND COSMOS

The suicide ghosts said: Listen to the roaring river, which doesn't care if you live or die. You don't matter to it at all. How curious that the same word, "matter," is the substance of the universe and also means: important, meaningful. Matter matters. We are not convinced by words. When you suggest that humans should derive their identities not from human society alone but from the substance of the universe, you may have walked into a trap. Life doesn't matter to matter. We don't deny that rocks can have value to humans, that river cobbles can be pretty; but what's the value of humans to rocks? The vast majority of the substance of the earth doesn't know that you are here. The earth quakes, and earth disguised as docile houses falls on humans and crushes you. In your laboratories you can assay the chemical values of sand, and in your observatories you can weigh the stars, but the sand and the stars never disclose anything about their value to themselves or your value to them. Value is a language they do not hear or speak. It exists only in the eyes and minds of animals. That's why humans needed gods with eyes and minds. You don't meet any gods of oblivious rock or blindly rampaging water. It wasn't just an error of perception that gods have awareness and wills, wasn't just because humans imagined that lightning bolts were thrown with the same intentionality with which humans throw spears or light campfires. It was because humans needed to be valued, and valued on a larger scale than they could provide for themselves within human society, valued for identities rooted in something stronger, in the cosmos. But the only way to achieve this was by turning the cosmos into society, a society of gods and humans. The gods fully understood human needs, for they too were lonely in an empty universe and created humans to keep them company and to proclaim that the gods have value, and when humans failed to worship the gods constantly, the gods felt empty and angry. Humans turned the universe into an admiration conspiracy. But there you lay, past midnight, listening to the roar of the pretty rocks that don't value your life, watching the pretty stars that with all their power cannot sustain themselves, the only gods you have, with fear your confessional.

Yet there were other ghosts in the Abyss. I saw the ghosts of the Ancestral Puebloans walking down to the creek and dipping pots

200

THE ABYSS

whose sherds I would find a thousand years later. I saw people digging up agave and piling up rocks for roasting it, rock piles still here and still charred today. The people who lived here faced a harder life than Puebloans living on more open and fertile plains, yet they chose to live here, for maybe two centuries; they chose to live. I heard the echoes of rituals that encouraged and celebrated living.

I saw the ghosts of the desert bighorn sheep that had been at home in the Abyss for ten thousand years. When humans showed up and took a lot of extra trouble to live, making pottery and houses and religious ceremonies, the bighorns thought it odd. Bighorn sheep never needed gods to tell them their lives mattered. They heard this message from inside, from somewhere deep and powerful. They heard the god of cells speaking, the god that in confused humans would disguise itself as the gods of rivers, rocks, and stars. The bighorns recognized this message being proclaimed all around them, by the frogs in the creek, by the squirrels and lizards, by the exclamation-mark birds, by the insects and even the flowers, especially by the newborn bighorn sheep.

There were still more ghosts here, the ghosts of rivers, rocks, and stars.

The river had risen up and given itself new shores and walked on the land and now was listening to the voice it had always had but never heard or heard from a rim a mile above as the sound of a distant and alien force—the river now was hearing itself being amplified off the cliffs and off the stranger walls of the human mind. Like a spirit returned to Earth with supernatural vision, the river was fascinated by its own substance and motion and glow, its formlessness and yet reliable forms, its currents and waves and tiny splashes, fascinated by how a river is a puzzle and how a kayak paddle is a key that can fit into the lock perfectly and unlock the river's secrets, fascinated by how humans didn't want to be locked out of the river when they could fit into it, immersing themselves in its power and beauty, how humans loved being here, on this river, in this canyon, on this Earth, in this puzzle of an existence.

Rocks put up a longer fight against dissolution than did living bodies, but finally they dissolved and disappeared. Now they were back, reincarnated as human bodies, trying to figure out if this was a punishment or promotion, dismayed by the flimsiness and briefness of living bodies, baffled by all their feelings, their fear of being hurt and dissolving. Hearing Granite Rapid, the ghost rocks imagined they were hearing rocks screaming at being dissolved, but no, only life screamed in that way. The rocks felt that land was natural and best and objected to being thrown into the annihilating river.

The stars died and went to Earth. The stars went to visit their grandchildren. The ghost stars stared up at their brethren but did not recognize them, stared at their new faces reflected in the river and did not recognize them or what they were doing here. Starlight slipped into the river like keys but got bent and could not open any secrets.

A few years later I flew off the rim and fell into the Abyss. I fell through the Kaibab Limestone full of fossils. I fell through the Toroweap. In the Coconino Sandstone I saw the sand dunes piled by ancient winds, and I felt those same winds roaming now, buffeting our helicopter. We fell through the Hermit and the Supai, the color of dried blood. Trees shot up like green fireworks. We were on our way to Crystal Rapid and my duty as whitewater safety researcher, even as the Guardian of Life. As we fell, I was trying to spot a car that had flown off the rim and into the Abyss a few weeks before, tumbling one thousand feet. But we were moving fast and I didn't spot anything.

The next time I went out the rim to the Abyss, I took binoculars. I spotted something too square, too shiny, a line of debris, a shattered windshield, a detached hood, leading to the car itself. I stared into the driver's seat, empty, except maybe of blood stains, blood that had also dripped into the canyon and become red flowers, supernovae mapping the distant beauty yet also the violent, unsacred geometry of space-time.

I plotted a line from the car upward to the rim and walked to that spot. I found a gap in the rocks and trees, and in the middle of it, a yucca with blades bent and sliced. I tapped the yucca and the whole plant wobbled, unnaturally loose. The philosophy of yuccas and cacti is that animals don't like to feel pain and won't touch knives

THE ABYSS

and needles. This philosophy had failed against a car, against humans actually eager to smash themselves against rock, rationally studying the best pathway and gunning their rational technology. I noticed that one yucca blade was peeled and dead, but another blade was healing itself. It was summoning all its cellular wisdom and powers to repel its invasion by death. It was arguing with the human who had decided that life was not worth living, that death was worth violent pain, that the canyon's geo-skeleton was as far as Earth should have gone, not into life, not into minds. But against an animal who could calculate the trajectories not just of cars but of the cosmos, calculate not just the orbits of rocks but the value of it all, what could a yucca say in reply?

Nausea

After Granite Rapid and a long day in which we ran a dozen major rapids, including Crystal, and had to figure out some unexpected trouble spots, I was beginning to feel more confident we'd get away with this whole trip. We camped amid a long riffle and had a big dinner and went to bed under a nearly full moon. But at 1:00 a.m. I awoke with a queasy stomach. As long as I lay on my back it was tolerable, but when I turned on my side, I felt nauseous. For me this was very unusual. Had the stress of the trip finally turned physical? I turned too much and nausea welled up and I crept to the river and threw up. After another hour of trying not to move, nausea took control again. I barely slept for the rest of the night. I lay there, trapped by my own body. I listened to the river, now an oppressive noise. I tried to remember if the NPS had any protocols for the correct place to vomit. I watched the stars moving too slow and too fast toward a very bad day to come.

In the morning someone asked me, "How are you?" and I lied, "Fine." But this was no ritual morning greeting. Several other people had become ill, and worse than me, with chills, headaches, and stomach pain. It looked like food poisoning. By the end of the day more people were sick, including two of our raft guides, too sick the next morning to continue rowing, and they had to let passengers row. When we ran into another raft trip we learned that they, too, had been hit by the same symptoms, which ruled out food poisoning and made it more likely that a norovirus was running wild. We had no shortage of medical expertise—four of our raft passengers were doctors—but they couldn't do anything. Sandra had been sicker than me that first night, but once again, as at Horn Creek, was determined to keep going.

A few miles below camp was Waltenberg Rapid, yet another rapid that was straightforward at higher flows but treacherous now, with a deep hole and not enough margin. One of our rafts fell into the hole and slammed hard, and one of the passengers, Teresa, lost her grip and flew forward into the metal frame, cutting a large gash in her forehead, the kind of impact that could break a neck. Blood poured down her face. Our doctors examined her and decided her neck was intact but might still have a significant injury—only a hospital could make sure and take the best measures, the sooner the better. We would call for a park helicopter and evacuate Teresa. The only way to summon a helicopter in that era was with a ground-to-air radio, and in this part of the canyon the only radios overhead were in jetliners miles overhead, with pilots who'd never imagined they'd get an emergency call from inside the Grand Canyon, who had to relay the call to flight controllers who had to relay it to the national park. We also needed to find a beach where a helicopter could land, and that took us another two miles. The guides spotted a jetliner and pointed the radio and called and called but never got any acknowledgment. Sometimes planes were visible only briefly before being eclipsed by the cliffs. Two guides climbed to a higher stratum to get better radio range. They called for hours and never got any answer. We hoped a message had gotten through, so we listened for a distant helicopter's *whomp, whomp,* but heard only silence.

While we waited, I climbed the slope. In a rock cavity, in a coffee can wrapped in a plastic bag, was the Journal of Tiki, actually Charlotte, who had died of Hodgkin's lymphoma at age nineteen. For a dying wish her parents had promised to take her down the Grand Canyon. She loved the idea but was too ill to go. After her death her aunt took her journal down the river and left it as a memorial.

THE ABYSS

Today I look at my face which is all broken out and I come
to the conclusion that I'm just another human and that life for
me isn't going to have me perfect and lovely. . . . It's just going
to be, if I live, a life of tragedies, love, sadness, the life that every
human faces. I'm going to know a lot about acceptance from my
sickness, how to accept going bald, being imperfect. . . . Perhaps it
will make me forget myself a little and give more to other people
which one has to do to live in this world. . . . For I have found
that if I don't love others I begin to be meaningless and I begin
to dislike myself. Then life becomes very weary and I halfway
don't want to go on.
But I can't completely let go.

With the radio battery almost dead, with evening coming, our
guides gave up calling for help. The doctors were monitoring Teresa,
watching for even a subtle weakness of sensation or motion, and they
decided she hadn't suffered neurological damage and didn't require
a hospital. But for her forehead wound, the longer treatment was
delayed, the likelier she would be left with a major scar.

More people were feeling sick. We made camp. Later we learned
that our call for help had gotten through but had gotten garbled,
and the rescue helicopter went to the wrong place, searched further,
found nothing, and gave up.

I asked where's God, in me or outside.
Do I believe in him.
Why are we here.
Do we have a purpose.
And I kept coming up with <u>Yes</u>,
<u>To live and to love</u>
Yes, we are to live and to go on to learn to love
It's an adventure, in many ways not pleasant.

A few years later, Tiki's journal disappeared. Someone stole it.

In the morning we decided not to try again for an evacuation.
Actually, Teresa decided. Given the choice between leaving the river

CANYON AND COSMOS

trip and having a lasting scar, she would accept the scar. I was surprised. She and her husband had been talking happily about going to Las Vegas after our trip, going to casinos and shows, which I considered a non sequitur for a canyon trip. Las Vegas glamour did not include facial scars, and Teresa was a pretty blond who seemed to be attending to her appearance more than most people on wilderness trips. But it seems the canyon, that long ugly scar, was telling her that Las Vegas glamour wasn't so ultimate.

We had lost a lot of time. We were only halfway through the canyon, with only four days to go, four days of thirty miles each, a major grind, no more time for hiking. That night, half of our group were too sick to show up for dinner.

INFLICTION

Why were we inflicting all this trouble on ourselves? River runners and other adventurers seldom pause to scrutinize their motivations, but our situation did prompt a long, floating discussion between Roy, me, and one of our raft guides, Lynn. Lynn had majored in psychology and written a paper on risk-taking, an autobiographical subject, for he was one of the first to kayak some of the most difficult drops in the California mountains. Roy was one of the pioneers of designing and flying hang gliders. After one week in which four of his hang glider friends died in four accidents, Roy went to a counselor to sort out his shock and motivations. Were they nothing but foolish, macho, adrenaline junkies? Lynn offered that kayaking was how he felt most alive, how it heightened mental and physical energies, uniting your mind and actions with the river's actions, a delightful marriage.

Yet the thrill of action is also felt by hunters and by soldiers killing enemies, hardly an honoring of life. The people who drove or jumped into the Abyss probably tasted some of the thrill of hang gliding—one suicide had said he wanted to stop being three-dimensional and to fly and thus talk with God. If thrills or simply mild pleasures are our justification for living, the suicides had run out of satisfactions and crossed the ledger into negatives, maybe depression. We could conclude that nature did not do the best job of designing human minds and bodies to handle the stress of living. Is nature indictable in a deeper way? Humans get a clearer look at the big equations of living than do other animals, and maybe suicides see more clearly than other humans that the big equations don't add up. For people committing suicide because of the failures of their social identities, we could blame society and not nature. In coming to the canyon, when they could have walked down the street and bought pills or a gun, they were asking nature to give their deaths more grandeur. Yet when Dale killed himself and killed his experience and love of the canyon, I could not avoid considering this a rejection of the entire natural order. Dale had greater authority on this, a stronger connection to break. If Dale was a priest of nature, he had denied his god.

When the Santa Fe Railway began bringing tourists to the canyon in 1901, a number of them, including prominent clergy, businessmen, journalists, and authors, reported feeling suicidal impulses. Henry Van Dyke was a national Christian leader and very popular poet, his poetry full of Christian praise, but the canyon triggered an existential vertigo in which he saw humans as "a conscious grain of sand / Lost in a desert of unconsciousness / Thirsting for God and mocked by his own thirst." Van Dyke felt "the fathomless abyss" urging him to kill himself. Harriet Monroe, the founder of *Poetry* magazine, felt a similar spiritual panic, seeing the canyon as the face of an alien universe in which humans were not welcome, and she "fought against the desperate temptation to fling myself down into that soft abyss. . . . Death itself would not be too rash an apology for my invasion." Monroe's circle of poets, including Carl Sandburg and Edgar Lee Masters, responded to her experience by writing major poems treating the Grand Canyon as an existential landscape for exploring God, nature, and human meanings. Henry Van Dyke's cousin John C. Van Dyke, a prominent nature writer, spent the first chapter of his 1920 book about the Grand Canyon addressing what he called "the terror of the abyss," asking "why before this most prodigious beauty of the world does one feel tempted to leap over the edge?"

Part of the answer was that many early visitors did not see beauty but a hideous denial of it. Another part was that this generation had been struggling with new scientific ideas about a huge, ancient universe full of inhuman forces and empty of God, and suddenly here it was, manifest in rock and emptiness that dwarfed humans.

John C. Van Dyke spent the rest of his book justifying why the Grand Canyon is indeed beautiful and inspiring, yet in the end he concluded: "The mystery that surrounds her should remain a mystery.... We opened our eyes upon the world with awe and we close them at the last groping our way in starry spaces. May it never cease!"

This first generation of visitors, who talked freely of existential vertigo, didn't actually kill themselves. Later generations of suicides never talked so philosophically about their impulses.

After too many people had driven off the rim at the Abyss, the National Park Service took dozens of small boulders and blocked off the easiest routes between the road and the rim. These chunks of Kaibab Limestone are loaded with fossils, which declare to today's life: *Stop, we couldn't avoid extinction but you have a choice. We worked so long and hard to create you, how dare you throw away all our work; we would have loved to have your consciousness to know ourselves and the world.*

Roy had moved on from designing hang gliders and was now designing parachutes for NASA sample-return probes to comets or asteroids, which might hold organic molecules that on Earth flew onward into life, into all kinds of probes.

Occasionally I scanned the water. I hadn't told anyone I was looking for a corpse. Sometimes I spotted suspicious shapes that turned out to be boulders. I saw many vague blurs I never figured out. The canyon offered many mysteries I could never figure out.

The Hospital

Room 346: Life was strong, but rock was weak.

Room 348: Rock was strong, but life was weak.

Intensive Care Unit #1: Life was strong, but rock betrayed it.

Room 346: I was visiting a friend, Julie, who had been severely injured by a fall in the canyon. Julie worked at Phantom Ranch, and before that as one of the mule wranglers. Phantom Ranch staffers, who have to hike in and out of the canyon, are avid hikers and like to try obscure routes. Julie was ascending a steep route and grabbed on to a rock, trusting her lift to it, when it broke off and she fell. When she woke up in the Flagstaff emergency room with broken ribs, pelvis, knee, and jaw, with two punctured lungs and missing teeth, the first thing she asked the doctor was how long it would be before she could ride a mule to Phantom Ranch.

Room 348: Some of Julie's fall was due to chance, to grabbing the wrong rock. Chance also decided that the person in the room next to her had also survived a canyon fall, but an intentional fall, a suicide attempt.

Intensive Care Unit #1: Gravity is very democratic: it treats all matter the same. It doesn't care whether people are falling on rocks or rocks are falling on people.

Rafters were camped three miles upstream from Phantom Ranch when an intense rainstorm lubricated the cliff above them and sent boulders falling and rolling onto their tents. Rhesa's pelvis was smashed, and she was bleeding badly and going into shock.

Room 346: The rock screamed, or echoed Julie's scream, as if thousands of tiny fossil mouths that had been silenced eons ago were upset that another brief chance to live was seconds from ending and that they were to blame; they had broken their promise of rock strength and broken off. The rock hadn't intended to murder anyone.

Room 348: The rocks said: With all our strength, we support you. You can stand on us forever and we will not fail you; you can build your house on us; you can quarry us and build great cities and temples from us. But the man heard the rocks saying: *Die. Join us in obliviousness. Being is easy if only you don't have to be human. We will knock all the confusion and pain out of you.* So he jumped. He was flying. He was free. The rocks rushed to embrace him.

Intensive Care Unit #1: They had wanted to go down "the Great Unknown," as John Wesley Powell had called it. They had found it. The river flowed without knowing where it was going. The rain fell

CANYON AND COSMOS

without knowing what it was doing. The boulders fell off the cliff without knowing what they were hitting. Millions of years of rocks falling unknowingly had built a canyon of unknowing. Humans felt compelled to know it. They came from far away and worked hard to know it and sometimes got the illusion they had gotten to know it, when its deepest secret was its unknowing. Rocks breaking off cliffs are supposed to remain unknown and unknowing. But humans seeking the unknown had placed themselves in the one unknowable spot and unknowable moment where the unknown became known, where the unknowing universe erupted as a scream.

Room 346: Julie told me that as she was falling she was shocked and angry at her foolishness. She was hiking alone on a route that, no matter how careful you were, was unsafe. She'd seen all the trouble canyon hikers got into, but she'd imagined she was immune: the canyon would never hurt her. As she fell she thought of Beverly, another Phantom Ranch staffer, twenty-six years old, who'd been climbing a shortcut five months before and fallen a hundred feet and died. Julie couldn't believe she was repeating Beverly's death.

Room 348: The would-be suicide stood outside Julie's door, saying, in a dazed way, he wanted to talk with her. A doctor was busy with Julie and told the suicide to come back later, but he just stood at the door a long time. He waited the way Death is always waiting at the door. Julie had felt Death close to her in the canyon, and with two punctured lungs she'd called out to make it go away. Even in the hospital, Death was right next door, stalking her. The doctor ordered Death to go away.

What did the suicide want to ask Julie? They had just shared the same, very rare experience of falling to your near death. Did they, as tradition claimed, see their whole lives flashing before their eyes? Did Julie fall out of the ordinary world and catch a glimpse of God? Or did she see absolutely nothing? Was Julie secretly relieved she was all done with nothing? Wasn't her carelessness almost a death wish? Or perhaps the suicide needed to confess to Julie that the moment he took flight he regretted it, he saw his own troubles in the right perspective for the first time, he saw the universe clearly for the first time and was horrified to be leaving it. Or maybe he wanted to ask about the injustice of someone who wanted to live getting smashed up while someone who wanted to die was walking around.

Intensive Care Unit #1: I could have introduced myself to Rhesa by saying I was a living being who was distressed that another being who shared this moment on Earth was hurting. But wouldn't that sound weird? So I said only that I'd worked for the kayaking school in the same Colorado town where she worked for a rafting company, and I'd been there only days ago and the cottonwoods were beautiful yellow and two river guides she knew had just had a baby, Kestrel Rose. Rhesa accepted me as a non-crazy person.

She recalled how to guard against ominous weather they had pitched their tents against the cliffs. In the middle of the night she was awakened by an explosion, the cliff cracking off. An instant later she was crushed.

The orderlies came to change Rhesa's sheets, and I stepped outside and talked with her mom, who cringed while Rhesa, being moved, was screaming.

Room 346: What binds us to this life? Perhaps this room held some clues.

Is it a mule? A wrangler friend had given Julie a mule doll, soft and cuddly, assuring infants that the world they have entered is soft and cuddly. Yet Julie knew well that real mules are not soft but tough and stinky and sometimes stubborn. She also knew the magic of looking into the big brown eyes of another shape of life and agreeing that going into the canyon is a worthwhile adventure, and that mules will safeguard human lives with every careful step.

Is it the flowers, their colors and scents and patterns and especially the stems and roots that knot them into the earth and funnel earth and water into their beauty? Is it the window view of a mountain turned gold with October aspens?

Is it the long row of cards above the bed, each a thread of caring that ties you to other lives?

Is it the human stories on the television, the enthrallment of a social animal in its own feelings and actions? Right now those stories

CANYON AND COSMOS

are incomprehensible, for Julie has turned off the sound so we can talk, but if the sound was turned up would those stories remain fundamentally incomprehensible?

Room 348: I wanted to ask him: why? I wanted to find out if he had now become a higher authority on life and death.

His door was open, but I knocked anyway. He didn't look up. With his teeth he was tearing at the bandage on his right arm. His left hand, too, was bandaged and thus of no help.

"Hi, how are you doing?"

He didn't look up.

"I was visiting Julie next door and I heard that someone else had, ah . . . fallen in the canyon. Where was it that you fell?"

"I didn't exactly fall," he said listlessly, not looking up.

I had asked a few people about his story, and one said he'd denied trying to kill himself. But his nurse told me: yes, he did.

"Oh, what happened?" I asked him.

"I was shot out of the barrel of a cannon." A cannon, a canyon.

He continued biting his bandage and tugging on it. "Help me get this off."

"I sort of doubt the doctors want you to take that off."

"Help me take this off."

Trying to make friends, I pointed out there was a piece of tape keeping the bandage wrapped and he'd have to remove the tape to loosen the bandage. He stared at the tape, bit it and peeled it loose. He worked the bandage loose, revealing a long line of stitches.

"I don't think the doctors want you to do that. You need to heal. You don't want to hurt yourself, do you? Why would you want to hurt yourself?"

He ignored me and stared at his stitches. He raised his arm to his mouth and started biting at his stitches. I went and told the nurses, two of whom rushed in, scolding him: "I guess you aren't any good at keeping promises."

Now he was angry at me, so I left.

Intensive Care Unit #1: Rhesa told me they were camped above

THE ABYSS

Zoroaster Rapid, and she had to be rowed through Zoroaster to get to Phantom Ranch.

I thought of how the prophet Zoroaster had tried to absorb polytheistic gods, such as the rain god Tishtrya, into one wholly benevolent god. In polytheism, rain gods were free to be capricious, bringing generous harvests for a while, then droughts and floods and misery. To make Tishtrya fit nature's reality, Zoroaster had to invent rain demons to battle Tishtrya and bring chaos. When chaos won, it was because humans had not worshipped Tishtrya enough—it wasn't God's fault if people got hit by rockslides. So let us praise Tishtrya for bringing rain, benevolent, holy rain that grows Hopi corn and Navajo sheep and desert wildflowers and that makes canyon rocks rise up and dance and roll.

The Hospital Cafeteria: Julie's mother, Cynthia, came close to saying that Julie's fall was the gods striking back against human hubris. Back home in Oklahoma, Cynthia worked for a cultural heritage program, interviewing Native Americans to save fading memories, including spiritual beliefs. Julie's father was of the Creek tribe, whose main ceremony, the green corn dance, defined the autumn harvest as the beginning of the year, requiring fasting and cleansing and other acts of humility before nature. Amid all the differences between Oklahoma tribes, Cynthia saw the common theme of humility before nature and its contrast with the oil-lubricated strike-it-rich mentality of Oklahoma whites.

I pointed out to her a Hopi who was a leader of the most traditional Hopi village. I explained how for the Hopis the Grand Canyon is sacred, the passageway of spirits to the afterlife, and some Hopis, including this Hopi, would not take one step down a Grand Canyon trail.

"I'm with him," agreed Cynthia. If Julie's "accident" wasn't a supernatural rebuke, it was at least a geological punishment. The canyon was a mass of collapsing rocks, and humans shouldn't be messing around in it. When Julie got back to the park, her wrangler friends were going to give her a ceremonial paddling for her hubris. Julie agreed she deserved it.

Room 348: When I got there the next morning, the suicide's room was empty. The nurses had posted a guard at his door all night and now moved him across the street. I didn't ask if "across the street" meant the psychiatric unit.

When he wasn't looking, I had looked at his medical charts. I saw values for all sorts of physiological data. I didn't see any line for "the value of human life." I didn't even see his name. In the end, as always, life and death remained a nameless mystery.

The Bunkhouse: We were celebrating the birth of mercy into the universe.

Lights flickered on the little Christmas tree. A fire flickered in the fireplace, in the living room of the Phantom Ranch employees dorm. Gifts and wrapping paper lay scattered about. Beverly was there, with 131 days left in her life. We were celebrating the world's redemption from The Fall.

Patti, a canyon mule wrangler well known on the cowboy poetry circuit, was playing her guitar and singing to us (I was included in the "us" as a friend). Then she started telling stories of her adventures with horses and rattlesnakes, horses and coyotes, horses terrified of bears. Terrified of bears? I looked around the room, at teddy bears dressed as Santa. These teddy bears were created right here by Betsy, a longtime Phantom Ranch employee whose room held a sewing machine and two hundred teddy bears. She'd made a video of teddy bears, with elaborate costumes and props, having canyon adventures, hiking, mule riding, and rafting.

Phantom Ranch was an appropriate place for teddy bears, for its original facilities were known as "Roosevelt Camp" after former president Teddy Roosevelt stayed there on his way to the North Rim to hunt mountain lions. For Teddy, the best thing about living was killing. After his frail, sickly childhood, Teddy became obsessed with macho posturing and big-game hunting. Yet on a bear hunt in Mississippi he'd refused to shoot a tied-up bear cub, saying this wouldn't be sporting. A cartoonist made this incident famous, and a toy company used Teddy's name for its new bear cub doll. Soon millions of children were falling asleep comforted by a cuddly bear. With one

THE ABYSS

act of mercy, Roosevelt had changed bears from ferocious monsters into symbols of a nature that was cute and loving.

The rock rolled, rolled away from the cave. A brilliant light shined forth. Angels sounded trumpets and sang joyously. The Fall had been conquered. Out of his tomb, like a bear from his den, a bear merely hibernating, Christ emerged. Christ was smiling happily. His big round brown eyes were full of mercy. Christ's brown fur was plush and cuddly. Christ's sewing machine stitches were perfect. He did not attempt to bite off his stitches. Christ held forth his paws and proclaimed: "I am the light of the world."

Or maybe we were celebrating the falling and rolling of a rock through space. When Christians decided that Jesus was God born as flesh, they wanted to celebrate Christ's birth, whose date was unknown, and they assigned it to the winter solstice and absorbed all its pagan symbolism of rebirth. It was a rock rolling toward greater light and life that pulsed our Christmas tree lights.

When Teddy Roosevelt organized his Rough Riders for the Spanish-American War, two Grand Canyon pioneers joined him. On the charge up Kettle Hill, which made Roosevelt a hero and soon president, Buckey O'Neill's macho posturing got him killed. Dan Hogan survived and returned to the canyon to work his copper mine. But his mine was 1,500 feet down a cliff and not worth its difficulties, and he abandoned it. Decades later, it turned out that the worthless gray rock he'd been dumping to get to his copper, dumping into the Horn Creek drainage, was the purest uranium ore in America. It was the link between humans and the forces of creation, the supernovae that had created many of the elements in human bodies, just as Jesus was the link between humans and the forces of creation.

One day, Grand Canyon rocks will fall from the sky, fall on cities, fall on capitols and cathedrals and libraries and hospitals. Grand Canyon rocks will repudiate the presumption of humans that the stones they had quarried and piled up into houses belonged to them and adorned them when they'd only been on loan from the forces of creation. Grand Canyon rocks will fall and turn cities into canyons large and small, revealing the unsuspected strata atop which humans had sought to extend one day further their fumbling, mapless lives and awkwardly sought to govern the matter that during and ever since the Big Bang had remained substantially chaotic and ungovernable. Grand Canyon rocks will become the instruments by which humans commit suicide. Humans usurped the authority to destroy the whole web of life, although perhaps it was life itself, with merely human hands, that was committing suicide, for life had failed to implant in humans a sufficient respect for itself and instead left humans desperate to hear the value of their lives, desperate to hear the forces of creation speaking to and reassuring them, and when they had listened too long and too lonely and heard only silence, the only way to make God speak was by pushing the holy button, making nature roar at last, the light of the world.

Christmas

Over Christmas, Dale was hospitalized for depression, suicidal depression, and placed on medication.

If Dale had been manic-depressive he likely would have been given lithium, one of only three elements created in the Big Bang, created by the power and harmonies of physics. Fourteen billion years later it turned out that lithium was highly effective at bringing harmonies to the brains of bipolar humans. No one is sure why. But there is poetic justice that the energies of creation, still spinning inside lithium atoms, can heal the disharmonies that have crept into creation.

Dale's medication certified that if Dale was thinking that the natural world was not good enough and that life was not worth living, these were erroneous thoughts, merely a symptom of mental illness. If Dale was normal he would be at a shopping mall, smiling at bouncy Christmas carols, being a good consumer, buying happiness. If anyone is sad over Christmas it must be because of another medical condition, seasonal affective disorder, resulting from a lack of sunlight. The cure for this is Santa Claus, a brilliant psychological strategy for defining winter not as deprivation but as festive and jolly. If anyone is feeling that the natural world is not enough and that what really matters is Love, the cure is Jesus Christ, who came to Earth to announce that the universe was born of love. Yet Jesus himself was careful not to be born of love—what a mess—but of a virgin, even as he claimed he was here to investigate the natural world, to inhabit a body and walk on the earth and eat fish and watch graceful birds adding to the beauty of Sea of Galilee sunsets, when all along, before he arrived, Jesus had decided that the natural world was not good enough and that life was not worth living and that he would overrule and exit that world by committing suicide on his cross.

THE FLAWED EARTH

The earth is flawed. The earth has flaws that go miles deep, irreparable flaws. Humans are constructed of flaws.

It was the middle of the night and everyone on my private trip was asleep. In a dozen dreams at once, houses or sidewalks began shaking and rumbling, and quickly everyone was dreaming the same dream of the canyon shaking and rumbling. Only it wasn't a dream. It was, we later learned, the strongest earthquake to hit Arizona in thirty-four years. The vibrating ground shook boulders off cliffs and sent them rolling down slopes, with much noise. In the dark we couldn't connect this noise with anything real, so the earthquake seemed doubly unreal.

A few miles upstream from us the earthquake woke up Navajo river guide Ray Interpreter, whose name recorded a mixing of eras and cultures. Ray was working on a geology degree in college, so he also spoke the language of the earth. The earthquake became a river quake and shook the raft on which he was sleeping, and he bolted up. Geology in action! Erosion yelling! The earth was speaking to Ray about another river far below, a river of magma, on which floated a continental raft on a long journey. He was feeling the power that had raised ancient seabeds to become land, the rock not deadness but the work of life, eons and oceans of life, the strength that outmatched all ice ages, warmings, asteroid impacts, volcanic outbursts, and continental shiftings, the strength that built the giant bell now being rung by the fire inside the earth.

The earth is flawed, and magma never forgets it and is always probing the flaws. In the western Grand Canyon magma found a long crack bisecting the canyon and poured upward and poured lava into the canyon and dammed the river. The river started probing the

CANYON AND COSMOS

lava for flaws and eventually ripped the dam apart, but for a million years lava boulders continued washing into the river and flawing it. The humans pulling to shore atop Lava Falls stare into the biggest flaw on the river and feel it probing their own flaws, mapping out their uncertainties and fears.

It happened that Ray Interpreter's trip arrived at Lava Falls as my trip was scouting. Our trips had been leapfrogging each other for a hundred miles, and last night we'd camped nearly together, a dozen miles upstream at the mouth of National Canyon. Looking at the people on Ray's rafts, I saw people with no legs. Another person had legs that barely worked, and his arms too flailed awkwardly. One man was blind. Other bodies were choking with cancer cells or other failures and would soon be dead. Other people spoke slowly and awkwardly. Tucked into their rafts were wheelchairs and walkers and crutches, none of which worked well on sandy and rocky beaches. The earth is flawed.

The organizer and leader of this trip was a longtime river guide, Jeffe, whose cells had betrayed him and become chaos guerillas. As he lay in the hospital waiting to die, lay in a chemo-stupor, he dreamed of the canyon he loved, dreamed of its strong cliffs and strong rapids and the strength he'd had there, and he wished he could see the canyon one final time and maybe truly appreciate it for the first time and let canyon walls and not bare hospital walls bid him goodbye. When Jeffe unexpectedly recovered, he thought of all the dying and maimed people he'd seen around him and how they'd never seen the canyon, how its strength and beauty would enrich their lives and their deaths. He persuaded the National Park Service and some rafting companies to take on the considerable effort and risk of taking the dying and disabled down the river. Jeffe called it the Jumping Mouse program, from a Sioux story.

Last evening as we'd been walking out of National Canyon and toward the beaches where our trips were camped, one of the Jumping Mouse organizers, Ann, was telling me about a woman on a previous trip who was dying of cancer and using the trip to say goodbye to living. Walking with us, and repeatedly interrupting, was a boy who'd found and dropped, somewhere on this massive outwash of rocks, a pretty rock with a fossil in it. He was eager to show it to me. "I know I can find it. It was here somewhere." There were tens of thousands of rocks here.

The dying woman had been alienated from her family and she'd come to live for years in a Hopi village and the Hopis became her family. When her raft trip stopped at the mouth of the Little Colorado River, up which was the Sipapu gateway, she got out some items she had carefully packed for a ceremony.

"It was a really cool fossil," said the boy. "I'll show it to you. You can help me find it." There were hundreds of thousands of rocks here.

The woman got out a little bottle of wine and a glass and poured the wine. *This is my blood you drink.* She got out some piki bread, a wafer-thin, rolled up, ash-infused corn bread the Hopis use as offerings at weddings and other ceremonies. *This is my body you eat.*

"You can help me find it," the boy insisted. I answered: "I don't know. There's a million rocks here."

Her ceremony was quite a mix-up of history. After the Spanish conquistadors conquered the Hopis and left priests there to torment them further, the Hopis had killed the priests and torn down their church and scattered its stones, and they had resisted Catholicism ever since. Her ceremony was also quite a mix-up of religious symbols, of earth spirituality and monotheism, confusing the Sipapu with God's ethical courtroom, confusing wine with rain, confusing a harvest offering with the body of Christ redeeming the disobedience of Adam and Eve. But perhaps her efforts were no more confused and hopeless than all the other symbols by which humans sought to deny death or at least give it a greater significance, the processions and rituals and prayers and sacrifices and feasts and pilings of fossils into pyramids and cathedrals.

"You can help me find it." He was picking up rocks, seemingly at random, and turning them over. "I know you'll like it. *You can help me find it.*" Okay, I was scanning the rocks for fossils. There were two million rocks here.

The woman lifted her glass of wine toward the cliffs, a mile of fossils, half a billion years of death, death so strong it had squeezed oceans of elaborate forms into formlessness and into rock over and over, death that never paused in half a billion years and now was sweeping through her on its race into the future. Yet death remained strong only because life remained even stronger, racing ahead of death, and now life had generated a species that refused to accept death and denied it in any way possible, however confused and hopeless, even if to find a voice as ultimate as death it had to speak through the mouths of gods.

"I know it's here somewhere." He was turning over more rocks. "I know I can find it. You can help me find it." There were ten million rocks here.

I was tempted to ask the boy about the night the earth shook and rumbled and everyone woke up in confusion. Was he, too, confused, or did he accept this as just another normality of this strange canyon?

If the boy saw Lava Falls clearly, he should have been confused. As we scouted, some confused people who had arrived before us did something confused and flipped a raft, sending two little heads bobbing through giant, confused waves.

Below Lava Falls we waited for the Jumping Mouse trip, and as one raft, filled to the brim with water, came around the corner, the fossil boy in it recognized me and assured me, with the same confidence he'd had about finding that fossil, "I wasn't afraid, I was never afraid." One of our rafters had said: "That's the scariest thing I've ever seen."

As I was walking toward National Canyon, across its rock fan, I came upon a man immersed in a pool of water. I noticed the cane beside him and something about his eyes. I said "Hello," and he answered, "Who's that?" I explained I was from another raft trip and we'd been mingling for days.

I looked into the pool, just large and deep enough for one person, likely dug out by the last flash flood down National Canyon. The clear water revealed a mosaic of rocks of every color from every stratum above, rocks tumbled downstream for miles, polished into fruit curves, embodying the water's smoothness and ripples, rocks fitted together artfully. The water's surface too described the cliffs vividly.

I complimented the man on the beauty of his little pool and its

THE ABYSS

rock mosaic. Beauty is in the aye of the beholder. Because the Grand Canyon is a famously visual experience, it's understandable to feel that a canyon trip is wasted on a blind person, but in elementary school I had learned, from a blind teacher, not to underestimate the blind. I told the man I bet he experienced the canyon differently than other people, maybe better. Maybe, he said. But he was shy about being probed by a disembodied voice, so I wished him well and moved on. Now I was free to use my imagination to see, yes, see the canyon as the blind might.

The canyon is a deep darkness, which is not the same thing as night, not a deprivation or an invitation to sleep. The canyon is not a parade of light and shadow and time but the timelessness of rock, all of it black. The canyon is not entertaining images but an elementary and mysterious presence. The canyon has no separation from the night sky. The moon-loving moths are just as mythic as the moon.

The canyon is a stone, and not one you picked up for its appearances but because deep fate wanted you to meet. It is rocks rounded like human faces, as smooth as a woman's face, as sandy as a man's face. You feel different textures and densities, almost colors. With your fingers you are hiking through all the strata, miles of rocks.

The canyon is a cliff face your hands read like braille, read better than other people can, maybe better than lizards.

The canyon is sand, soft and loose. When you walk on a beach, the canyon assumes your own shape. You have made the canyon deeper.

The canyon is air lush with the scents of soil, water, and especially life. You are a coyote and you can follow invisible trails.

The canyon is sounds. It's water dripping from a cliff, the murmur of a riffle, the growing thunder of a rapid, the rain splashing. It's the sounds of the wind, a swaying bush, a peregrine falcon swooping and calling. It's a wonderland of cliff amplifications.

The canyon is an animal, human words that conjure the darkness into the shape of a bighorn sheep on a ledge.

The canyon is a message from a woman who lived a thousand years ago. It's a pottery sherd that, once whole, carried water or stored seeds and still holds corrugated groves formed by her fingernails, into which

CANYON AND COSMOS

your own fingernails fit perfectly, allowing you to read her mind, to feel the struggle and the beauty she knew here.

The canyon is motion, the strange new motion of a river and a raft, speeding up and slowing down, bouncing and spinning and tilting.

The canyon is a little pool of water reflecting the canyon. As your toes enter the water, the canyon vibrates as in an earthquake. You immerse yourself in the canyon, stratum by stratum, up to your knees, your hips, your chest. You are the river and the canyon is parting for you. The cliffs reshape themselves to embrace and portray you. With only your head above water, your body is the canyon and your breath is making the canyon pulsate, giving it life.

If you entered this pool at night, it would be a pool full of stars. As your toes touch the surface, the stars vibrate as if from a supernova. Up to your knees, up to your waist, up to your chest you immerse yourself in the sky, which embraces you. Your body is the cosmos. Your breathing is the breathing of the cosmos. Even your heartbeat vibrates the stars. You melt into the beauty of the night. You had always been the night.

As nineteenth-century science increasingly encroached on religion, as physical forces took over the roles long performed by an all-seeing God, a metaphor arose: the universe is blind. Blindly acting forces rule the universe; blind chance drives evolution; blind fate dominates human lives. For an animal dependent on vision, the worst thing you could say about the universe wasn't that it was deaf or mute or stupid but that it was blind. Christians may not have made the sun into a god, but they still began the universe with light and symbolized God with light and crowned Jesus and angels with light and made light and darkness symbols of good and evil. In a blind universe no one was steering the lightning bolts toward justice or regulating human behavior or curing sickness or adding up the rewards in your account. A blind universe did have advantages, for comets were no longer omens and your child's death wasn't a punishment for your sins, but the blindness metaphor wasn't the best consolation for grieving parents.

The blindness metaphor was true, mostly. The universe is far more generous at generating light than vision. The universe was

THE ABYSS

born as light and became galaxies pouring out light, but only crater eyes received it. Light stirred gas planets into wind but not attention. For the longest time, the universe performed its wonders for no one. But here and there, stars sent out umbilical cords of light and birthed symbols of themselves, as round as stars, star gates, black holes hungry for light, turning light into recognition of light. On Earth, light-hungry trees lured animals away from the ground and away from scent; trees like hands reaching skyward sent humans to mountaintop observatories to sniff out the vaguest and oldest trackways of the cosmos.

Yet the universe's blindness remained so powerful that it sometimes overflowed into humans and brought trouble and searching. Blind Homer, acutely aware of the punishments of blindness, pivoted the *Odyssey* on Odysseus escaping the cyclops Polyphemus by blinding him, only to so anger Polyphemus's father Poseidon, the sea god, that Poseidon inflicted years of ordeals on Odysseus. Still, Homer "saw" ocean waves full of poetry. John Milton, troubled that his blindness might be a punishment from God, troubled by the plagues and other injustices around him, looked through Satan's eyes at the complexities of good and evil in hopes of justifying the ways of God. The Norse made Odin give up one eye to drink from a cosmic spring and gain its wisdom.

I once hiked down the Bright Angel Trail—the brightness of angels was another sign of the dependence of humans on light and vision—with a blind friend. Keith had been a Grand Canyon ranger for years and had worked at Phantom Ranch and hiked the trails to it many times. He was not entirely blind; he could distinguish basic shapes and light patterns; but the flat earth held enough tricks to trip him occasionally. On the trail Keith's pattern recognition warned him of mysteries that needed to be probed more thoroughly, first with his cane, then with his foot. Occasionally he misread the ground and stumbled. One time, not reading a gravel section, he slipped and went down. I helped him back up and looked at his leg scratches. I wasn't offering him any guidance; I knew him too well. Other hikers looked at Keith like he was crazy, but of course he couldn't read their faces.

CANYON AND COSMOS

One hot August day, in the shade shelter beside the river at Lees Ferry, I was hanging out with Lonnie, who the next morning would become the first blind person to kayak the canyon. Or try to. Lonnie was telling me how he'd come to be here, how he'd been blinded in a hunting accident, hit in the face by a shotgun blast; the pellets were still embedded. Lonnie retreated to his Indiana farm, where everyone told him he couldn't do anything ever again, until one day his five-year-old daughter, Bug, offered to guide him on his riding mower around their weed-choked lawn, and this went so well that he "never looked back." A navy submarine veteran, Lonnie liked boats and water, even being underwater, and soon he got a kayak and was practicing rolls in his scummy pond. He was at Lees Ferry with a group of veterans, some disabled. Other kayakers would guide him, yelling out instructions. I didn't tell Lonnie that even sighted kayakers often can't react fast enough and accurately enough to the waves smashing into them. Lonnie's kayaking experience was seriously thin, and I couldn't decide if his happy-go-lucky attitude came from a strong personality or dreadful naïveté.

I was intrigued by the idea of kayaking blind, and tried to imagine it. The river is a strange, dense mass of motions. Your kayak becomes a spinning, confused compass needle. Turn, turn, turn, the swirls and boils and eddy lines won't even tell you which way is downstream. Other kayakers can "read water," but for you hitting a rock is a surprise. The river talks to you, yells at you, in a foreign language. Rapids are full of mysterious powers, kabbalistic waves, and you begin to sense, as did Helen Keller as pump water poured over her hands, that the river holds deeper meanings.

It turned out that Lonnie did well. It turned out that he had both a forceful personality and a corny sense of humor that got him through the inevitable punishments. In Lava Falls he flipped and rolled up and flipped again and was ripped out of his boat so violently that he wondered if his legs had been broken, and his boat cartwheeled down the rapid as he was swept along underwater.

The earth is flawed. Yet even the flaw that created Lava Falls also left a crack from which warm water splashed down the cliff right below the cold rapid, a warm shower into which people could place their faces and hair and feel refreshed. If neither rock nor water knew anything about human bodies or had any intention of honoring human needs, if there was a flaw in the earth that left people without legs or vision or mental quickness or much time for living, this did not diminish the gift of a warm spring on human faces.

One by one the rafts came around the bend and into the warm spring eddy, including Ray Interpreter's raft. There was no drumming. When Ray became the first Native American to row Africa's Zambezi River, tribesmen drummed the message down the river, "Here comes the red man," and people gathered to look at him. There was no drumming now, unless it was the splash of his oars, splashes nearly as quiet as the splash with which, eleven days later, Ray would fall into the river and drown. At his funeral, conducted partly in the Navajo language, a woman in a wheelchair, who had been on the Jumping Mouse trip, sang some hymns, sang them to God, or maybe to the flawed earth, or maybe to a Big Bang so intense it had blinded everyone.

THE NEW YEAR

Afterward, some of Dale's friends wondered if the antidepressants he started over Christmas actually backfired, making him more prone to suicide. Clinical research suggested this could happen. Yet maybe this idea was also a way of denying that Dale had cast a valid judgment against the world: he wasn't thinking straight, he didn't know what he was doing.

Dale could be a rational observer of his own irrationality. The naturalist who was fascinated by every twist of a canyon cave was fascinated by the odd workings of the human mind. But this didn't mean he could escape his dark confinement. We agreed that depression had a logic of its own, against which you can't argue logically. When his work season neared, he was unable to function and could not do the job that gave his life a purpose.

Some of Dale's coworkers were reluctant to talk with him about depression and suicide, but our river trip, in which Dale had trusted his safety to me, had defined our relationship differently, so we talked candidly. I asked him to promise that if he ever felt suicidal he would give me a call, and he promised. He broke his promise.

SHILOH BELLS

Long before he could see it, John Wesley Powell heard it, the loudest roar he had heard in the Grand Canyon, made more ominous because the canyon's red rocks were disappearing behind blackness. With his geological imagination Powell saw a river of lava flowing into the canyon and river: "What a conflict of water and fire there must have been here! ... What a seething and boiling of the waters; what clouds of steam rolled into the heavens!"

Powell and his men climbed through the lava boulders and stared into Lava Falls, except that it wasn't Lava Falls yet. For Powell, as for the first Native Americans to see the canyon, nothing had a name. For Powell the canyon's mystery was magnified by the continual mystery of what lay around the next bend, a mystery not merely geographical but personal, the mystery of whether the canyon would let Powell live or kill him.

For the first humans to see the canyon, the canyon was simply itself, unpolluted by human desires. Yet humans began hiding the canyon behind their hopes and fears, their desires for passage and food and shelter and meaning. Native Americans turned the canyon into gods they could talk to about human needs. Like Adam, Powell began naming things, names that fooled subsequent river runners into thinking they knew the canyon, left them unable to see it as naked rock, pure mystery.

Powell stood there feeling the roar of Lava Falls with his whole body. He had not heard a roar like this in seven years, at least not as sound, but sometimes he heard it as pain in the stump of his right arm, through which the Battle of Shiloh refused to stop.

The cannons Powell commanded at Shiloh were an invasion of geological chaos into the biological world. Their iron had swirled deep inside Earth for billions of years, generating magnetic fields and aurorae, being squeezed out as volcanoes and being buried again, while above it Earth was transforming itself from naked rock into a

THE ABYSS

swarm of life. Now life had summoned the iron out of the ground and given it shape but only for life to inflict on itself geological shapelessness, the same chaotic force that through volcanoes set cities afire and that through earthquakes set cities tumbling. Life looked at its eons of work and said: it is bad, and I shall smite it with a rod of iron.

The cannonballs flying toward Powell had a further origin. The Confederate states lacked generous iron deposits, and to gather enough iron the Confederates put out a call for bells, plantation bells, schoolhouse bells, courthouse bells, church bells that could be melted down and recast into cannons and cannonballs. General P. G. T. Beauregard wasn't the only Confederate who paused over the idea of turning church bells into cannons, but he decided it would be righteous because it would "rebuke with a tongue of fire the vandals who in this war have polluted God's altar."

From all over the South, the bells came.

From city churches with steeples that dominated skylines, from log cabin churches in Ozarks backwoods, from Louisiana swamp churches whose bells had informed alligators and snakes and mosquitoes that the universe is a realm of benevolence and immortality, from coastal churches whose bells had rung louder than buoy articulations of the worried sea, the bells came. The preachers themselves got out their ladders and brought down the bells and felt righteous about it. Bells that every Sunday with pure musical beauty had called people to praise God, bells that at Christmas had celebrated the enfleshing of divinity, bells that at weddings and baptisms had proclaimed that more than crude biological impulses wanted life to continue, bells that had promised resurrection and heaven to corpses in coffins being carried to their pits, the bells came down. From Baptist churches and Presbyterian churches and Catholic churches whose preachers had ranted against one another's errors and destinies in Hell, the bells came together and harmonized.

Of course, this meant that across hundreds of cities and towns and country miles, God fell silent. Leaving only the swooshing of the wind and the rattle of merchant wagons.

For this re-bellion the bells were remolded into rather similar shapes, hollow iron tubes longer than wide and open at the end,

CANYON AND COSMOS

with a bulged outer rim, with a round iron ball inside that made noise, except that now this ball didn't remain bouncing inside but flew outward. Church bells had always been as hollow as the hollow men who needed to be filled.

Now this hollowness spread out like angels' wings over the Shiloh battlefield and filled it with hollowness, digging thousands of craters, digging hollowness into human bodies and faces. Now God called out over the battlefield, a warrior god with a voice deeper than bells, roaring against his imperfect and disobedient creation. God's roar tore apart bodies, breaking loose arms and legs and heads, smashing skulls and brains into splinters and pulp and blood, spraying blood into the faces of the living. God declared the fragility of human bodies, the disloyalty of blood, their readiness to melt back into the mud from which they had been molded.

God tore off the arm of John Wesley Powell, leaving him with a hollow sleeve for life, leaving him with pain, leaving him unable to swim or row a boat or climb a cliff, as if God was trying to prevent him from going into the Grand Canyon and seeing the evidence for geological eras and biological evolution, prevent him from making a deposition about deposition, prevent him from spreading the deepest hollowness on Earth into the minds of humans, prevent him from dragging God out of the human-scale Genesis story and forcing Him to fit into the dinosaur tracks of a much older and vaster cosmos.

The cannonballs flew with mathematical precision, always writing out the same equations of force, weight, angle of ascent, gravity, air resistance, and distance. The cannons could have been clipped from the illustration in Newton's *Principia* showing a cannon firing a cannonball with varying velocities until with a sufficient velocity it became a moon in orbit around Earth. Some of the iron in these cannons and cannonballs had indeed performed astronomical orbits, as part of asteroids that knew their place in the solar system until Earth diverted them. The asteroids plunged into Earth's oxygen and began burning up and falling apart and roaring. The asteroids plunged into Earth's water and set off tsunamis. The asteroids plunged into the ground and set off explosions that darkened and chilled the world and exterminated millions of species in pain and chaos. From the perfect heavens, without oxygen or water to corrupt them into rust, the iron angels had fallen. The cannonballs falling now were reenacting their fall as asteroids, gouging holes in the ground and in flesh; life had chosen to inflict asteroids upon itself. Out of hollow life's yearning for identity and value, life was inflicting the hollowness of space upon itself. The iron was trying to reunite itself with the pure iron god in the center of Earth, but the messy biological world was getting in the way.

This cannon iron had also killed stars. As stars synthesized progressively heavier atomic elements, they began forging iron and built an iron core, a giant cannonball, but unlike all the elements before it, iron refused to ignite and fuel a star's continuation, and eventually the iron became too massive to support its own weight and it collapsed and started the rest of the star collapsing on it, triggering a supernova that blew the star apart and fried nearby planets, raining iron on planets full of life, even life for which that star had been a god.

Iron that had helped extinguish planets of life now flew from cannon supernovae and fell on a battlefield and blew apart bodies that had placed supernovae atop Christmas trees and sung carols to them. Iron that had directed the sinews of stars now contracted inside muscles to wave cloth stars and to run frantically in fear and rage. Iron that had flowed in stars now raced in blood, blood now gushing onto the ground, gushing out of biological orbits and into creeks it stained red, where it reenlisted in the armies of gravity and hydrology. What a strange journey this iron had made, from ironclad astronomical laws to the chaos and doubt of Earth, from creating matter to tearing it apart, from creating light to yearning to find meaning in light. It had become the iron of great telescopes and of stained-glass windows, of warrior face paint and of cemetery gates that would open wide on the day of resurrection and let the dead of Shiloh parade gloriously on the sunny and confident earth.

And yet, the star-destroying fury of supernovae was also a creative force. In an instant, supernovae synthesized the elements stars could not create, the elements heavier than iron, and spread them throughout space, where they might eventually congregate into new stars and into planets and red canyon cliffs beautiful with sunrise.

John Wesley Powell, an ardent abolitionist, might have said that the violence at Shiloh was a creative violence, forging a better cosmic order, breaking the chains of millions of slaves, shedding the European feudalism erroneously transplanted to America, replacing it with the democratic equality and respect America was supposed to have. When Powell saw the Grand Canyon, he saw not just geological and biological evolution but a cosmic order that endorsed human progress, not just scientific and technological progress but moral and societal progress. His cannon and his cañon advocated the same message.

General Grant was invading the South, unloading his army from the Tennessee River onto a landing with a log cabin church named with the biblical word for peace. Shiloh. The Confederates surprise-attacked at dawn, Sunday dawn, pushing the Union forces back through farm fields and orchards and dense woods, but some Union forces, including John Wesley Powell's artillery battery, took a stand and, with intense fighting, held off the Confederate attack.

At dawn all the leaves in the woods around the Shiloh church awoke, all the cells went to work to sustain life, employing all the reliable skills they had inherited from billions of years of cells before them. Through birds, cells sang to greet the dawn and another day of life. In rabbits and squirrels, in ants and hornets, cells loyally continued their ancient duty. In every human, every cell was declaring the value of life.

Why then, against the imperative in every cell, were thousands of humans rushing toward death? Why were cells indistinguishable from one another so eager to declare other cells to be worthless and destroy them?

To the Illinois farm boys in Powell's battery, their cannon volleys hitting the Confederate line reminded them of how storm winds toss rows of wheat. Even across a hundred yards, Powell and his men saw the red geysers erupting on faces and chests and legs and genitals, the heads and arms flying off, the bodies reeling backward and falling, the groping of men now blind, the open mouths that must be screaming but could not be heard over the battle roar.

Sometimes it took awhile for all the cells in a body to get the bad

THE ABYSS

news. The brains knew it at once: men saw their leg ripped off or their stomach ripped open, saw the blood gushing out, and they knew they were going to die. But in other parts of the body, cells continued their normal work, weaving the future, oblivious to the futility of it, just as they had never known the individual they'd been serving, just as their man had always been oblivious about them. But now men stared in astonishment at their bodies ripped open, at the muscles and bones and intestines that had always been hidden within them. The intricate molecules rushing on intricate schedules, the ever-growing architectures, the cellular universe had labored for a man, had loaned their skills to his purposes and pleasures in exchange for his fulfillment of their purpose, their dwelling safely within his body for a while and being passed on to dwell within his oblivious daughters and grandsons for a while before moving ever onward.

But now the news was spreading: the cells had been betrayed. Cells waited for nutrients that weren't coming. One cellular stoppage cascaded into others, until some final cell in stupid isolation learned that the universe had collapsed.

The cells had no idea why they were dying, what strange order of events had led their bodies to rush into fields swarming with bullets and cannonballs. The cells had never thought of themselves as Union cells or Confederate cells. The cells in Powell's right arm didn't know they were giving the signal to destroy other cells just like themselves.

In a magnification of the obliviousness of cells, hundreds of horses were running loose on the battlefield, many wounded and dying, and they had no idea why. Horses lived in a universe whose basic physics consisted of starting or stopping, turning right or left, moving faster or slower, leaving the barn and returning to the barn, carrying a person into town or pulling a plow across the field, and being fed by a friendly person. It made no difference to horses whether the friendly person was an Iowa farmer or a Georgia plantation owner or a black man with whip scars on his back. Now men commanded them to rush toward a huge, incomprehensible noise, to serve unknown, contradictory stories, and they ran and erupted and dragged intestines through the dirt and fell and screamed and died.

Did any of these men see what they were losing or killing? Did any of them see that what was rushing toward them was only themselves, the same cells, the same human body, the same life in a universe with little life, the same rare consciousness? Each body embodied the same human family tree, thousands of past species, billions of awakenings to dawns, and life's long learning how to live. Each body embodied the flowings of rivers and continents and the massive work of the stars. Each face glowed with the light of the Big Bang and of mystery.

But no, the only thing these men saw were the masks the universe wears in human society, the masks and uniforms that wholly define human value and that sometimes contradict one another. These strangers so much hated being contradicted and so little valued life that they were ready to destroy their own lives by rushing into a swarm of bullets and cannonballs for the chance to inflict their hatred on mere masks.

By what authority did these men assume the right to forever cancel other lives? If a human life is a soul implanted by God, were they not defying God and risking his wrath? In what courtroom was John Wesley Powell's command arm the gavel of what judgment? After thousands of years of prophets proclaiming and philosophers arguing, is this in the end our only way of deciding the value of human life, this storm of hating, maddened, rushing, terrified men destroying what they claim to value?

Listen: the drums are beating. Look: a Confederate brigade is lining up. Watch closely now: maybe we can learn something about the value of human life.

This brigade has already charged the Union line once, and they've seen other brigades charging it and seen men falling by the hundreds and seen the futility of it, and their commander has protested the futility of it to his general, but they are lining up to charge again, knowing they will die, die to accomplish nothing, knowing they are committing mass suicide.

Yet look: they are advancing into the open. They are staring into the black holes of John Wesley Powell's cannons; they are seeing futility. Yet doesn't this futility look rather familiar? Haven't they seen it while sweeping the dust from their shops every morning, while pulling the weeds from their fields every week, while sitting down to dinner to hold hunger at bay for a few more hours? Haven't they seen it while shaving beards yet again and admiring their youthful vigor in the mirror, yet seeing signs of aging that weren't there last year and recognizing that their bodies are destined on their own course toward wrinkles and frailty and death and they are powerless to stop it? Aren't these soldiers seeing only the same futility that had seemed remote, spread through thousands of future days, suddenly compressed into the present moment, looming, given the specific shapes of these woods and fields, this roar and smoke? Weren't these soldiers seeing only the enigma of death finally manifesting itself as measurable iron and lead? Wasn't their submission to futility now simply the recognition that futility would consume them eventually anyway? If you are going to die anyway, wasn't it better to die on a glorious battlefield than in an old man's piss-stained bed?

But wait, the cells are saying *STOP!* Surely we can blame this battle on the childish excitement of drums and bugles and waving flags. Surely with their deeper voice the cells will call out and stop this suicide. Or can we avoid blaming the cells themselves? These are the same cells that don't know their own bodies, that labor blindly for eons, that futilely continue ticking in a dying body. Aren't these soldiers simply conglomerations of stupid cells? Or are these soldiers the minds with which cells can finally see themselves clearly and pass rational judgment on the value of their labor? Listen: the drums are beating. Look: the men are charging, the men are committing suicide.

As men lay dying and their minds were flickering out, what was their final vision of the universe? Perhaps they saw one blade of grass, really looked at a blade of grass for the first time since they were children, and were astonished by it. Perhaps they looked into their torn body and only now realized the marvel of bodies. Perhaps they saw a distant house or wife or child, loved now more than ever. Was their final verdict on the universe regret for having to leave it? Or was the horror of this battlefield an illusion-stripped revelation of the horror of existence?

For John Wesley Powell, the value of human life was being decided by his right arm, giving the command to fire. He felt like Thor, the Norse god whose thunder-striking arm upheld cosmic order and

CANYON AND COSMOS

justice. Yet suddenly, with a shock and a scream, the gods stole from Powell the arm that held the value to decide the value of human life. What sacrifice was this, on what altar to what god? Thunder was branded into him forever, but what was this thunder saying? Now Powell could not convince anyone, including himself, of the value of his own life. With the aching stump of his omnipotence, he would submit himself to the thunder of the Colorado River, desperate for the verdict it would deliver upon him.

Suppose you are walking across a field and see a rock, crudely shaped, merely sitting there, serving no purpose. Now suppose you find a pocket watch. The watch's complexity indicates the existence of a skilled watchmaker, who designed it for a purpose. Then suppose you find a living creature with far more design and function than a pocket watch. Such a design proves the existence of a designer, who designed creatures for a purpose, a benevolent purpose. This benevolent designer must be God.

Thus reasoned William Paley in his 1822 *Natural Theology*, the most famous exposition of the argument from design, which insisted that the order of nature proves the existence and benevolence of God. People might be questioning the Bible's logic and miracles, but anyone could see the message in nature. Isaac Newton had proven that God governed not through sporadic magic or arbitrary force but through natural law, a thoroughly rational and reliable clockwork universe. If kings had justified their arbitrary force as licensed by a tyrannical God, the deists, including Thomas Jefferson, argued that a God of rational laws mandated a government of reason and law. Through reason, humans could discern the ethical laws God had implanted in nature, and discern it as "self-evident" that all humans have value and equal rights.

So here, at last, is our proof of the value of human life: in a field stands a horse on whom sits a Confederate brigadier general, in whose pocket ticks a watch. This watch has ticked off the Sunday hours in which the Confederate army has done its best to destroy the earthly government of cosmic rationality and equality, ticked off the pain and final moments of the 23,746 men wounded and killed at Shiloh because the value of human life may not have been so self-evident after all. It ticked on at the command of the coiled natural laws of Newton's clockwork universe, ticked with the same predictability with which a firing cannon lurched backward with an equal and opposite force, proving Newton's second law of motion, the same predictability with which bodies, struck by cannonballs, lurched backward with an equal and opposite force and divided their equations into severed and flying heads and arms. The watch did not pause an instant when its owner, Brigadier General Jason Lycurgus Compson II, had his arm blown off.

William Faulkner does not tell us exactly how Brigadier General Compson loses his arm at Shiloh, only that Shiloh sets in motion the decline of the Compson family, which Faulkner chronicled in *The Sound and the Fury*. Of the novel's four chapters, three are titled "April 6," "April 7," and "April 8," the dates of the two days of the Battle of Shiloh and the Confederate retreat to the town where Faulkner was born thirty-five years later, where two thousand Confederate wounded were gathered and hundreds were buried. William Faulkner's grandfather, John Wesley Falkner, told his grandsons stories about their great-grandfather Colonel William Falkner in the Civil War, how he led a charge that captured four Union cannons. Grandson William was awed by the colonel's silver pocket watch.

If William Faulkner isn't going to tell us how Brigadier General Compson lost his arm at Shiloh, we are free to imagine that he lost it to one of John Wesley Powell's cannonballs, perhaps the one fired when Powell raised his right arm to command his cannons to fire but lost his own arm.

Brigadier General Compson returned from the war a defeated man, and he passed this defeat to his descendants. His alcoholic son sold their final parcel of land to send grandson Quentin to Harvard, their final hope of family redemption. One morning Quentin awoke in his Harvard dorm room and "I was in time again," hearing his grandfather's pocket watch ticking, the same ticking that had measured thousands of deaths at Shiloh and sent John Wesley Powell toward the Colorado River to prove his life still held value. Quentin heard the enormity of time, "down the long and lonely light-rays you might see Jesus walking." But in the watch his father had given him out of family continuation and hope, Quentin saw only wreckage. He taped the watch against the dresser and broke off the glass and grabbed the hands and twisted them off and dropped them into the ashtray: "The watch ticked on. I turned the face up, the blank dial with little wheels clicking and clicking behind it, not knowing any better." Signifying nothing. It was a clockwork universe that had no message for humans. That night, Quentin drowned himself in the river.

It was not the Colorado River, into which John Wesley Powell, a broken watch with only one hand, had stared and seen the enormity of time and the enormity of chaos yet glimpsed not just himself triumphing in a river but a river embodied and ticking and triumphing in him. Quentin tied flatirons to his ankles and turned himself into a mere volume of gravity, sent himself flying toward the stars reflected in the river, propelled by the iron forged in the chaos of stars, iron that had killed stars in supernovae and that flew from stars with all the elements of future human bodies and that at Shiloh had flown again to debate the value of those assembled elements. Quentin fell as if denying the whole course of cosmic evolution. With a splash that sent perfect circles spreading upon the river and rocking the stars like supernovae, Quentin disappeared.

Eighteen years later, on April 8, the day after Shiloh, which this year also happened to be Easter Sunday, Quentin's mute idiot brother Benjy was being driven in the Compsons' battered surrey to the cemetery to lay a flower at Quentin's grave, as he did every Sunday. Every Sunday when they approached the town square with its statue of a Confederate soldier who "gazed with empty eyes," the surrey went around the right side of the square. But today the driver turned to the left, and when Benjy saw the Confederate soldier on his right rather than on his left where it had always been, he bellowed in shock and horror.

Benjy saw and heard everything happening around him but comprehended none of it. Events happened for no reason. People and things appeared and disappeared with no purpose. The present and the past blurred together. Clocks didn't add up. Church bells meant nothing. Benjy was continually baffled and upset. When people and things disappeared, Benjy whimpered, moaned, bellowed. "It was nothing. Just sound. It might have been all time and injustice and sorrow become vocal for one instant by a conjunction of planets."

Yet perhaps Benjy had a deeper and truer vision of reality than those who judged him incompetent, people who might have a name for every object and a sequence for every event and who might recognize patterns in human families and in human society and in the solar system, yet who deceived themselves into imagining that the universe is comprehensible, when in truth events happen and lives appear and disappear for no ultimate reason, when Shakespeare agreed with Benjy that life "is a tale told by an idiot, full of sound and fury, signifying nothing."

Sound and fury reigned over the Shiloh battlefield.

When darkness fell and the guns fell silent, soldiers on both sides heard the moans and pleas of a thousand wounded and dying men left between the lines, a cacophony of pain, men begging for help, crying to their mothers and sweethearts, crying to God, getting no answer. They heard wounded and crazed horses staggering about and stepping on wounded men and crushing their chests and skulls. They heard the gleeful snorting of hogs feasting on human intestines and brains—they loved brains for their salt, the molecules of which had recorded the images and sounds of the battle, the final surges of pain and horror. The hogs tried to eat the intestines of men still alive, who thrashed and screamed. The stench of death filled the air. Among the wounded soldiers lay the stars, the stars on the flags with which both sides had sought to invoke for themselves a greater strength and glory.

Late at night, thunder began, rain began, heavy and relentless, turning the ground into mud and streams. A strong wind chilled bodies and broke off bullet-cracked tree branches and crashed them onto the wounded. Mixed with the thunder was the roar of Union gun boats on the river, roaring all night, often hitting the abandoned wounded, keeping everyone else awake all night. Not just in the minds of the delirious, the thunder and the guns blended together into the same hideous assault. The thunder roared like God's cannons. When the thunder rolled away, the night was again filled with bellowing.

Every few minutes a lightning bolt revealed the battlefield, the smashed trees and cannons, the roaming horses and hogs, the wounded and the dead, all flickering for an instant and disappearing into the darkness again. Lightning exploded with primordial power. Lightning exploded like the echo of the Big Bang. Not just for the delirious, the echo of the Big Bang felt like a hideous assault, as if the Big Bang itself had been a cannonball exploding and the entire universe was its wreckage, as if all the stars were the fallen flags of defeat, as if the Big Bang itself had wounded the men flickering and bellowing on this battlefield. The mad swarming of men at Shiloh was a continuation of the mad swarming of particles in the Big Bang, a continuation of the swarming of particles inside stars, a continuation of the swarming of nebulae and thunderclouds and volcanoes and oceans and rivers, all searching for something, trying to build something, now swarming to try to assay the value not just of humans but of this entire long swarming yet perhaps finding only that this entire planet is a wasteland, that an entire universe of lives have been left wounded, abandoned, without help, bewildered, bellowing like Benjy, flickering into existence for an instant and back out, writhing with the agony of not knowing why, of not knowing what it all signified.

In Lava Falls, John Wesley Powell felt the sound and the fury, an echo of Shiloh, of the thunder and rain and pain that had bombarded him all night as his arm lay outside the hospital tent in a pile of hundreds of arms and legs still bleeding, and he continued trying to lift his vanished arm (but not as men beside him continued trying to lift their vanished arms to pray for mercy) and he anguished that his life or his future or his value was over.

As Powell had gazed up at the canyon's limestone cliffs for weeks, he sometimes heard the rain. For hundreds of millions of years, it had rained. All over the world, it had rained. To the bottom of the ocean had rained arms and legs and heads and shells and stems and blood. Death upon death, corpse upon corpse, inch upon inch, ton upon ton, layer upon layer, the raining corpses had built a mile of rock. If the canyon was a record of life, it was also a record of death.

Corpses rained down because they had been allocated only a small and brief talent for living and they had exhausted it. Corpses rained down because they'd been deprived of the sunlight or nutrients or something else they needed. Corpses rained down because creatures had infected or devoured them, battling over who deserved to live. Creatures fell into a mile-deep grave ticking time mindlessly.

Into the mass grave of all evolution, into the open wound of the earth, into the unweighable wreckage of the past, into a river's devouring of tombstones, into the voiceless rapids and the raging silence, into the clockwork parade of giant shadows, into the hall of carnival mirrors, John Wesley Powell had come, just as Walter Kirschbaum would come a century later from the same endless swarming and mad confusion, to find some deeper significance.

What will the canyon grave tell him? With what voice will it speak? With what shattered arms will it draw his image? It does seem to be trying to say something. When he speaks, the grave cliffs echo him. When he touches rock, it speaks of strength. When he walks, the pebbles speak of the time rolling onward within him. When he sits at night, the moon and the campfire draw his image on the still unfolding, ever creating moments.

Atop Lava Falls, Powell stands aghast. Sometimes under stress he forgets and thinks he is himself again, his old, whole self, and like Brigadier General Compson pointing with his vanished arm to this vanished land and identity, Powell points into the sound and fury of Lava Falls, points with nothing, points at nothing, signifying nothing.

MYSTERIES OF THE DEAD

If things had gone as planned, I would have been staying with Dale at his North Rim home on the day he killed himself. He'd given me a long-standing invitation, and I was going to take him up on it before joining a North Rim archaeology project. If I'd been there or on my way, he might have changed his plans. Twelve days before his suicide he made out his last will, and on the day I would have been there he left it, along with three letters, on his kitchen table. People who saw him in those twelve days said he'd seemed happier than he'd been in a long time. No one recognized why.

As fate decided, another friend of mine drowned in a strange kayaking accident on a California river, and I decided to remain in Flagstaff for his memorial service. I tried to call Dale, and left a message that I'd visit him after the archaeology project.

Even if Dale hadn't just killed himself and left me wondering about my role or failures in it, exploring archaeological ruins sets you thinking about the dead. You feel ghosts here. Some of the people who had lived here were buried somewhere nearby. I wondered if any of them had killed themselves. Did they feel abandoned by their climate, their bodies, their friends, their gods?

A FLAWED CREATION

Nathan fell.

He fell toward a stone floor that lay atop hundreds of feet of limestone formed by millions of years of falling, by millions of bodies falling to the seabed every minute, tiny lives and tiny deaths unnoticed by anyone, falling, falling, falling, falling, adding up to hardness. Bodies fell at the command of the same gravity that tore down mountains and dug canyons and bent tectonic plates. No body could refuse to fall. Almost all were destined for obliteration, leaving not even fossils, merging into an anonymous mass. Nathan was merely one more body falling toward the seabed.

And yet, out of limestone had emerged the stones of the pyramids and kivas and cathedrals through which life declared that life was not gravity or rain or rivers or mountains or canyons but something new, more defiant than rock.

There wasn't an instant to hesitate. If I had hesitated, Nathan would have hit the hard stone floor. Without arms to absorb his impact, he likely would have landed on his head. Yet now fate was handing Nathan a rebate. If Nathan's weight was normal for his age, he might have fallen too fast for me to stop him. His lightness meant that he teetered for an instant, that gravity was slower to grab him. But there wasn't an instant for questioning, for philosophy or religion, even for personal sympathies for a child or his parents. There was only an instant to act or not act, and perhaps my acting emerged from a deeper impulse, from the devotion of cells and heartbeats to themselves, from the impulse that kept life rising out of the seabed, rising against the fall of bodies, rising into pyramids and kivas and cathedrals. I jumped forward and reached out. I was denying not just the fall of a child but the Fall of Man, denying the claim of pyramids and kivas and cathedrals that humans are embedded in a logical and nurturing and moral and immortal order—I was refusing to protect God's reputation by blaspheming innocent children, blaming them for their own suffering. If I was only another body waiting my turn to fall, only a body with my own instant left to live, and the only power I had was compassion, even human compassion so flawed it could not stop people from lighting Auschwitz crematorium fires or taking karmic consolation from them, then this was all I had to give. I was also denying that this is a universe of nothing but force, of gravity and rivers and flames. It does offer compassion.

Just barely, awkwardly, I caught a flawed and frightened child. I embraced a flawed creation.

I lifted Nathan back onto the desk. He prostrated on the wood, away from the edge, buried his head in his arms and sobbed, shook with sobs. Milo put his arms around his son and tried to comfort him.

When it was time for me to leave, I reached out my hand to Nathan, careful to select the correct hand, but instead of taking a handshake Nathan puckered his lips. I leaned forward, offered my cheek, and he kissed me. This was Nathan's answer to the Fall. Lacking a muscular arm—like the one with which God reached out and tapped Adam and declared that creation is good—and trapped in a creation of flawed, weak, diseased, confused, doomed bodies, this was still his answer.

THE GOOD RANGER

Even in the end, Dale was a good ranger. I suspect he considered the idea of sneaking off to a canyon cranny where no one would ever find him, where he would melt his elements back into the rocks and junipers and ravens. But he had seen how missing persons or rim jumpers imposed a great deal of trouble and risk on park rangers.

CANYON AND COSMOS

He wasn't going to do this to his friends. He wanted more order, even as chaos triumphed. So he arranged his letters and his will on his table, went into his garage, rigged a rope, as if delivering frontier justice, and jumped, not far, yet very far.

Dale's house was right on the canyon rim, with a view for which people came from all over the world, at great expense, to enjoy for a few hours and remember forever.

None of Dale's friends saw this coming. With Katy, I compared our last conversations with Dale, and she, too, had seen no warnings. She had once talked with Dale about suicide, and Dale had disowned it and even joked about it, and she decided it could not happen.

A helicopter flew Dale across the canyon, as if ceremonially, over the shrines, over the Colorado River, over the Puebloan ruins. The ravens flew with him.

LEAVING THIS EARTH

Nathan lived to age twenty-three, exceeding expectations. If enjoyment of living counted for anything, he willed himself to a longer life. He loved music and animals and comedy and jokes, loved making others laugh. Though his body didn't allow much athleticism, bowling requires only one arm and he became good at it. He became the batboy for his high school baseball team. His head never grew any hair, but Monica liked to think of it as "the most beautiful, shiny head to ever exist."

Milo didn't live to see any of this. Less than three years after our cookouts and Nathan's "fall," Milo was dead. He did not drive off Hopi Point. He developed non-Hodgkin's lymphoma, nearly the same disease that killed Tiki. It did not have any of the logic of uranium poisoning or Auschwitz karma. He went home to his parents to die, at age thirty-seven. Even knowing he was doomed, he did not drive off Hopi Point.

LOST

As the river miles passed and we got farther from Ron's last known location, I grew less hopeful I would find him. It was a relief.

I never did find him.

The next spring, some hikers were wandering a few hundred feet

THE ABYSS

off the Tonto Trail near Grapevine Creek and found Ron's body, mostly his skeleton. He had never been "in the river" at all. Why had the search team decided he was in the river? Perhaps after their exhaustive search, this had seemed the most logical conclusion. But if some randomly roaming hikers found him without trying, the official search was far from complete. Perhaps with the Secret Service demanding results and the National Park Service unable to deliver, bureaucratic self-protection prompted them to offer a safe result.

I was left wondering if Ron really had killed himself. In the first story I heard, Ron's skeleton was hidden behind a boulder, in the shade, still wearing boots. The National Park Service helicoptered it out. They brought in Ron's hiking partner to identify his clothing and backpack, but he couldn't. It took a medical examiner to identify Ron, but it was too late to determine a cause of death. In the second story I heard, Ron had fallen some fifty feet. Perhaps it was an accident, perhaps not. You can also choose to die by simply sitting down behind a boulder and giving up.

I would never know the truth. I would never know the truth about all sorts of things about living and dying.

ASHES

Ashes to ashes. But in between, ashes have a chance to say something, black ashes can gather in Blacktail Canyon and sing.

At Dale's memorial service on the North Rim, his cremation ashes sat in an urn. It perplexed me. The night before, in an aspen forest beneath a mountain, with millions of aspen leaves fluttering, with a bagpiper and his soulful lament slowly, slowly emerging from the aspen whiteness, I had pondered another urn with the cremation ashes of another friend with whom I'd done a Grand Canyon river trip, who had not chosen to die.

Dugald was a professional photographer and working on an adventure story assignment in California. He was an expert kayaker but was ambushed by a vastly improbable sinkhole in the bedrock, down which water was dropping into a cave, leaving no clear warning on the surface. Dugald passed over some swirling water and instantly and inexplicably was pulled backward and downward, getting wedged in, water pounding over him with tremendous force. He struggled, he was strong, but he could not get out of his boat, and he drowned. Dugald would have given anything to avoid this mistake and go on living. Dale had thrown his life away.

Inside Dugald's urn was a river, a river that had invaded every source and every outlet of a human will and washed it away and left only carbon and calcium unrecognizable and unrecognizing, carbon and calcium waiting with the non-will of a river, the non-will of mountains and stars, waiting for hands still inhabited by will to take them and release them where they belonged, into a river, the Colorado River, which meant that the river inside the urn also would be released, would be allowed to be itself again, even celebrated for being itself, forgiven for carrying from the mountains to the wombing ocean the sediments that didn't know or care how often they had been alive and had enacted through magic carbon masks their brief protest against being only ashes yet again.

I stayed at Dugald's memorial service until midnight, trying to stay warm near a bonfire transforming wood into ashes but also into light, stayed partly because I was afraid to go home and sit there alone thinking of the morning and the long drive to the North Rim to meet another urn full of ashes but not any river, full of the non-will of rivers and mountains but also a will that had decided to reject the tiny and brief universe of wills, leaving me no river to blame or to forgive for only being itself, leaving questions no river or mountain could answer or even ask, leaving me with a storm of conflicting feelings.

As I brooded over the contrast between Dale and someone who had fought for his life, I shared my feelings with Katy and she protested: he did fight, he fought for years. This was true, I had seen it. The human mind too holds rivers, powerful and full of strange traps; depression is a river with only vague rules for navigating it, overwhelming your will and intelligence and bravery, as strong and treacherous as Horn Creek Rapid.

At both memorial services people talked about the lives lost but said almost nothing to try to make sense of their deaths.

When we finished sharing memories of Dale, we went out to a

THE ABYSS

quiet spot on the canyon rim. We were led by Dale's two brothers and his father. They opened the urn and each lifted a handful of ashes and held it over the rim and let go, at least physically.

One of Dale's brothers turned toward the rest of us and held out the open urn. No one moved. We had supposed we were here only to watch. Hesitantly, one person stepped forward, dug into the urn, and stepped to the rim.

The sky was still wounded from this morning's storm, the winds still strong and confused, the clouds still swirling. When someone released Dale's ashes, some were caught on an updraft and lifted and swirled into a black wisp that quickly vanished. For one person after another, the ashes both spilled into the canyon and sprang upward, sprang into chaos and into a pattern, dancing, quickly vanishing. One especially strong gust created a black tornado. Were we watching a soul being released and ascending to heaven? Or a black soul revealing it had always contained the chaos to which it was now returning? Once, at the very moment ashes were spiraling upward, a raven shot from below the rim and glided upward right alongside the ashes, as black as them, watching them, feathers vibrating to the same currents as them, and when the ashes disappeared the raven completed their soaring into the sky.

I was the last person to step forward. I did like the idea of helping Dale merge into a place he loved. But I remained horrified by his suicide, and to help him complete his rejection of living seemed to make me an accomplice. Perhaps I was feeling I had already stained my hands.

But none of us had had a chance to say goodbye to Dale, and this was it. I looked into the urn and, silently, started to tell Dale I was sorry this had happened to him—but then I saw that this wasn't him, or anyone you could talk with. The urn held only a landscape, a strange, unrecognizable, desolate lunar landscape, gray and black with flecks of white, the aftermath of some volcanic eruption or forest fire. I couldn't comprehend it. I couldn't comprehend the connection between this landscape and the person I had known. Or not known. I had journeyed all the way through the canyon with this landscape, relating to it as a person when all the time it had secretly

245

been this, a moon following its own mysterious pathway. I couldn't understand how such dust could rise up and live, soar up as a raven, stand here as myself. I couldn't understand what kind of powers could send these ashes into life and send them back into ashes. Amid all the universe's reliable orbits and quantum jumps, how are such decisions made? As if this was a mirror I was looking into, a black mirror, a skull mirror, this formlessness was dissolving my mind into formlessness, into unrecognition. Stunned with mystery, with both horror and beauty, I reached into the urn.

I reached in and felt powder softness and hard bits of bone and skull. I grasped firmly and removed my hand carefully, trying not to spill anything—but wasn't I being silly to be so careful after what Dale had wasted? I stepped to the edge, and unlike the others who had reached straight out and allowed the wind to seize control, allowed confusion and restlessness to have the last word, I kneeled down and put my hand against the cliff. I'd had enough of confusion and restlessness and was aiming toward peace.

I let go. The ashes and bone and skull fell and whirled and spread out until I could no longer see them, for a human cannot see far or clearly into such depths. I thought of these ashes becoming soil and trees, lizards and ravens, further river journeys.

We sat on the rim for a couple of hours and played guitar, with blackened fingernails, and sang some of Dale's favorite songs, including songs we'd sung in Blacktail Canyon, in the night, with the full moon looking for us, with frogs joining us. To that billion years of lost time, lost by an unworried rock hourglass, we had added more lost time, felt the sands of our own lives draining into the wind.

The Raven

THE RAVEN

On the boundary between solid earth and shapeless water, the boundary between order and chaos, between life and death, there stood a raven. In my turbocharged vision and mind, the raven became a deeper vision.

Right behind the raven the Colorado River was collapsing. The river's strong, deep smoothness was wavering in the middle and unraveling around the edges, and then the entire river fell and crashed and fought itself and roared.

The black rock on which the raven stood had once flowed as strongly and plunged as chaotically as the river. It had poured red hot out of the volcano on the rim above and spilled into the canyon and dammed the river, until the river overflowed the dam and ripped it into rubble, through which the ghost volcano still churns the river today, and churns human minds.

Lava Falls may be the most famous rapid in the world. It's always the culmination of a Grand Canyon trip, and not just because it waits at mile 179, but because it's the hardest challenge. It's not the longest rapid, but a longer rapid also gives boaters more time to maneuver around obstacles. Lava Falls is a quick plunge into chaos. The topography funnels too much power into too small and obstructed a passage. Right at the top, the middle third of the rapid is effectively a waterfall, which is why this is Lava Falls and not Lava Rapid. This waterfall creates a deep and long hole, "the ledge hole," the most frightening hole most boaters have ever seen anywhere. To the right of the ledge hole the river drops into a huge trough, the "V-wave," pushing

CANYON AND COSMOS

both ways at once, breaking unpredictably, tilting rafts steeply upward and maybe sideways too, maybe too steeply. Beyond the V-wave is a series of chaotic lateral waves, pushing unpredictably. The current aims at a protruding huge, rough, lava boulder, given nicknames like "dead man's rock" and "the cheese grater." All this violence adds up to the deepest roar in the canyon. Even the most experienced professional guides approach Lava Falls with dread, for they've heard all the stories, maybe left a story or two. River guides enjoy testing their skills against a river, but they don't like turning their fate over to dumb luck, and at Lava Falls the ratio of dumb luck to skill is higher than anywhere else.

Your chances of getting through Lava Falls depend heavily on starting out at the right spot, with the right angle and direction and speed. At some river levels the best run is very close to the ledge hole, and being off by even three feet can drop you into the ledge hole for an instant flip and violent pounding. Even when you miss the ledge hole, three feet can make a big difference. The moment you enter Lava Falls it overpowers you, leaving you unable to change course; the best you can do is adjust the angle of your raft to fit the next wave.

Lava Falls is even scarier because it's a thoroughly blind drop. As you approach, all you see is the river disappearing, a horizon line—a black hole event horizon—with nothing behind it but a few geysers of spray. You can't see the ledge hole or wave trains. When you reach the horizon line you suddenly see the entire rapid, a truly awesome sight, but if you've made a mistake about your location, it's too late. Of the many tricks Lava Falls plays on you, the worst is giving you a false sense of knowing what you are doing. When you scout Lava Falls you climb onto a volcanic ridge two stories high, and from here you see the rapid spread out like a map. But beginners often fail to study the subtle location markers above the drop, the riffles and swirls and lines of bubbles, and how they connect with the dynamics in the rapid. When people get back in their boats and approach the falls, they are often shocked to find they cannot correlate the view in front of them with the map they saw when scouting. They see a geyser but can't figure out where it's coming from. They panic. Boaters should be powering their way into the rapid, building momentum to better

THE RAVEN

overcome the push of the waves, but it takes a lot of nerve to rush into a trajectory when you can't see where you are going, and even experienced guides hesitate too long.

By the third time I kayaked the canyon I had realized that the most important thing about running Lava Falls is studying the subtleties above it. On this trip the other kayakers, none of whom had been here before, rushed ahead to scout, but I stopped and studied the top riffles and eddies. I stepped slowly and watched how the perspective changed. I was focusing on one wave in particular, an unmistakable long, gently sloping wave with a tip that crested and lapsed every few seconds. On my previous visit I had figured out that if I aligned my kayak with the righthand crest I would be in the best location and angle for dropping into the rapid. I had also told myself that if I started paddling hard here, angling leftward, I'd have the best chance of avoiding the V-wave, but I'd frozen and waited until I came to the horizon line and got swept into the V-wave's tyrannosaurus jaws and slammed over. Now I was reassured to find that my marker wave was still here, still pointing correctly, and I studied it intensely, watching it pulse, making sure it wouldn't disappear on me.

Rationally, I had decided that Lava Falls was manageable. There was zero chance I was going to fall into the ledge hole. As ugly as the bottom boulder looked, the lateral waves were pushing away from it, and the water surging onto it was quickly surging off, a phenomenon with the comforting name of "pillowing." The worst thing that was going to happen to me was that the V-wave would knock me silly, but this had happened twice before and I'd waited out the further turbulence and rolled up. This was my reassuring worst-case strategy: hold your breath for twelve seconds and roll.

But humans are not entirely rational creatures, even in calm circumstances. Lava Falls filled me with a dread that registered itself throughout the body, in heart, breath, muscles, stomach, bladder, sweat glands, and fidgeting.

From the scouting ridge, the trip leader sent the two ace kayakers to go first, with everyone watching them. Everyone was dismayed when they headed straight for the ledge hole. They had not realized how much they needed to veer to the right to find the wave train. Everyone yelled and waved and pointed frantically. At the last moment, the two veered rightward and avoided disaster.

I had not waited to watch them run and headed back to the boats alone, wanting to have a further study of my favorite wave. Launching alone, without the usual camaraderie, did feel lonely, even more so because I couldn't know that the first two kayakers were waiting down below in case I needed help.

My nervousness was not the social tension of facing a school test or work assignment where failure meant a loss of respect. It was the tension of cells recognizing a threat to their existence and sounding an alarm and rallying their deepest resources. Rationally I was still insisting that this was not a matter of life and death—*calm down*—but bodily I was not listening. The ancient primitive animal deep inside was stirring and taking over.

I felt unreal, first because I could not entirely remember what I was doing here, why it had seemed a fine idea to kayak the Grand Canyon. Lava Falls, too, now seemed unreal, just too bizarre.

I felt surreal. The first ability I needed now was perception, highly acute perception, the perception to recognize one little wave and be *right here, absolutely here, not three inches off, not ten degrees off.* Then I needed to correctly perceive every wave coming at me very fast. My mind was firing my senses into full jet thrust, especially my sense of place, of presence. My heightened alertness seemed to be applying itself more broadly, for now the river seemed to be flowing with added intensity, and the boulders around me seemed blacker and sharper. When I picked up my kayak paddle, its wood-grain waves were more vivid, downright amazing.

I didn't have time to dwell on any of this. I slid into the water.

From the right shore protrudes a series of volcanic boulders and slabs, at first obstructing your view of the right side of the approach. This can distort boaters' sense of location, sending them straight ahead, toward the ledge hole, when they need to be veering rightward. As you pass these boulders, the approach to the rapid is progressively revealed.

I wasn't looking at the last boulder in this row, but I knew I had to be mindful of it, for behind it the current spun into a strong eddy.

THE RAVEN

I was watching for my *be here* wave, and I finally spotted it, but of course it looked different from here, so I studied it urgently, its shape and cresting, but then it vanished—I was alarmed—but it pulsed again, right on time, right in the one essential place. *Be here, absolutely here, completely here.*

As I was approaching the last protruding boulder, something odd happened. The boulder moved. Not the whole boulder, just part of it. I couldn't afford to look—*be here, completely here*—and overrode this as a perceptual error, but then the boulder moved again and took on a shape and a face and black eyes staring at me. Standing on the edge of the boulder, blending perfectly with its blackness, was a raven.

My searchlight of visual and mental intensity was intercepted by this raven, as if by a lightning rod. I was intensely surprised, intensely astonished. I saw the raven with a greater intensity than I'd just seen the river and boulders, with more intensity than I'd seen almost anything in my life.

No time, no time for my mind to dredge up cultural files and wrap them around the raven. It was not a symbol of anything, not ominous, not shamanic, not poetic. It was not wildlife or matter. It was not aesthetic, not beautiful or ugly or sublime. It had no connections with the Grand Canyon. It was simply itself, intensely itself, naked strangeness.

The raven was not only intensely shaped, intensely black, and intensely eyed, it was intensely *here, absolutely here, completely here.* Intensely present. Improbably real. Impossibly existent. The raven's first and supreme quality was simply that it existed. Out of the realities that usually disguise it, existence was bursting forth and revealing itself. I was being ambushed by existence itself. It was staring right into me. I was absolutely awed.

Adding to the raven's strangeness was that it refused to act as ravens should. It was allowing me to come much closer than a raven on land ever would, at last only three feet apart, staring into each other's eyes, each other's depths.

Suddenly Lava Falls seemed even more astonishing.

I lined up on my trustworthy wave but once again failed to accelerate and entered too far right and was swept into the V-wave, and it firehosed me in both directions, dumb-luck failed to flip me, but it spun me rightward and shot me rightward onto an eddy line and I flipped, quickly rolled up, going backward, turned forward, ran straight into the bottom boulder—even as wild boils levitated me I was fascinated by how the facets of the boils matched the lava facets in the boulder—and the boils dropped me past the boulder and steeply into the eddy behind it and I flipped again. Rolling up, gasping for air, I stared at the backside of the boulder and the waves roaring past it.

Lava Falls wrenched me far away from my vision of the raven, but I wanted to hold on to it. As I waited and drifted now, and as we paddled away later, I scanned the boulders and the sky for that raven, any raven.

Two months later I returned to Lava Falls. My memory of the raven remained strong. A mile before the falls I began scanning the boulders and cliffs for ravens, but all this black, angular rock made it a poor place for spotting ravens.

When we landed and headed along the trail to the scouting ridge, I stopped and looked at the boulder where the raven had stood. No raven there now. I tried to be more rational about why the raven had been here before. Ravens will land on unattended rafts and inspect them for loose food, but from this boulder the rafts weren't visible. Had the raven landed here to get a drink of water? There were many far more likely spots for that. Why had the raven allowed me to approach so close? Perhaps it was as surprised as I was. Ravens have lots of curiosity, so ravens are likely fascinated by these weird humans careening through Lava Falls—but this was a poor theater seat for watching that. I couldn't see any logical reason for the raven to be here. This was only right. I don't know much about visions, but I knew that visions are not logical.

Approaching Lava Falls this time, I felt all the same tension, and approaching the raven rock I was ready for awe, but the raven wasn't there and I felt abandoned. I stared at the boulder anyway but nothing happened.

Once again I collided with the V-wave and got knocked over, and before I could even get ready to roll, some dumb-luck wave flipped me right side up again.

The next year I returned to Lava Falls three times, and with my few undistracted minutes I watched for ravens, and when I spotted one I concentrated on it.

On one of those trips I had unexpected hours on the beach below Lava Falls, as we waited for a helicopter. This trip included a paddle raft, which six people propel with paddles and a captain steers from the back. As our paddle raft plunged into Lava Falls, our captain somehow broke or maybe dislocated his wrist. As we waited on the beach below, usually a place of celebration, the midwestern Catholic priest who'd been on the paddle raft told me he'd felt the captain's hand knocking the back of his head and this must be how the hand got broken. I doubted this, but the priest seemed determined to be guilty. A few days before, after a hike requiring the hardest climbing he'd ever done, we'd had a conversation about human frailty. He said that our river trip spoke to him of human frailty, for at any moment any of us could die from falling, drowning, or a heart attack. He dealt with human frailty almost daily at a hospital, ministering to the injured and sick and dying. Often he was confronted with questions about God, including angry accusations. He told me he often didn't know how to answer, for he really didn't understand human suffering. When he died he would be able to talk with God and finally learn the answers.

Ontological mystery can break out anywhere, in many forms.

I wandered up the beach toward the falls. I scanned the boulders and the sky for ravens. When I spotted one gliding along the cliffs high above, so small, I was disappointed. It was only an ordinary raven, flying in an ordinary way. Only an ordinary raven above an ordinary rapid in an ordinary canyon on an ordinary planet. I was ordinary too. Still, I found ordinary ravens fascinating, for like many canyon lovers I saw ravens as the canyon's personification: the canyon is made of sky, and ravens are thoroughly at home here, living adeptly in a rough place where humans struggle. But now I wanted ravens to be more than ordinary, to glow with the surrealistically real energy that had struck atop Lava Falls. And if ravens did personify the canyon, was the Lava Falls raven offering some sort of message from the canyon itself?

Of course, I knew that merely seeing a raven wasn't likely to trigger the same experience, which had arisen not from the raven but from my own mind, under unique and powerful circumstances, with surprise a large factor. I knew that standing here safe and relaxed was mind-miles away from where I'd been. Yet I was reluctant to admit that this experience depended on Lava Falls or any danger or stress. I wanted to find my way back to that vision, to find its sources within myself, to cultivate it or spark it. I knew that a raven was not essential for this, that I should be able to have the same vision with a wren or a rock. This vision shouldn't require a Grand Canyon but should be available in any meadow or town. Yet my experience had been so strong it had fixated me on ravens and Lava Falls. Perhaps I'd felt cheated, being offered a vision for a few seconds only to have it ripped away by other priorities.

Lava Falls hadn't been necessary when, at age twelve, I noticed I was alive, noticed it with the force of a revelation. It was odd I hadn't noticed this before. Perhaps I'd been too busy playing games, watching TV, interacting with people and animals, figuring out how the world works. Perhaps the world was so full of novelties that it took me awhile to notice that I too was a novelty. I suddenly saw myself as if from afar, as a strange body and mind, a presence erroneously taken for granted. I'd never heard any friends mentioning the strangeness of being alive. Even adults seemed too busy to notice. Occasionally, in no predictable pattern, I'd have a flare-up of wonder. I saw a gothic clocktower lit up against the night sky, a tower I'd seen numerous times, sometimes seen as only the time regulating my day, but now it glowed not just with light but with vivid, inexplicable, almost supernatural strangeness. I would never again see it quite like that, but I knew it was possible, knew that it held a secret.

As I explored the adult world, I was disappointed to find that such moments of ontological wonder seemed rare. Novels and movies took human existence and bodies for granted and dwelled on only the complications of human life. Religion insisted it accounted for

THE RAVEN

everything, and so did science in its own way. But every so often I'd perceive the profound strangeness of things.

Super-surrealistic Lava Falls had offered me the strongest vision I'd ever had, then instantly ripped it away. I didn't want to let go so finally. But it wasn't easy to get back there, and when I was there physically it wasn't for long and I was heavily distracted. There was a "trail" from the rim to Lava Falls, but I'd heard horror stories about it and I couldn't interest anyone else in going, and anyway perhaps this needed to be a solitary pursuit. I'd just have to let go.

A few years later, destiny called. Actually, it was ranger Linda calling.

Years of researching the impacts of Glen Canyon Dam on the downstream environment were finished, and the findings generated a surprise proposal. The dam had trapped sediments in Lake Powell and eliminated the annual spring floods that deposited new sediments throughout the Grand Canyon, and for decades the beaches had been melting away, removing plant and wildlife habitats and camping options for humans. The research had included some pioneering hydrology, especially on how sandbars get built or eroded, and it suggested that unusually high river flows would dredge sediments out of the riverbed and deposit them on beaches. While the dam had cut off the canyon's main supply of sediments, side streams and flash floods had been banking sediment on the riverbed for decades, enough for a major restoration of beaches. An artificial flood of 45,000 cfs, several times normal, had now been scheduled. It would last one week. Dozens of scientists would be stationed along the river to study every aspect of the flood. The NPS wanted to continue its study of how different river levels impacted river trips. Linda wanted to know if I'd be willing to hike the "trail" into Lava Falls and spend a week there, all alone. She said this apologetically.

I confessed I had never hiked the Lava Falls trail. Linda said she hadn't either and didn't know what to tell me. She did relate a recent horror story of a park ranger who had tried the hike, which ended with a park search-and-rescue helicopter. This was not quite the reassurance I was hoping for. But I could not refuse the call of destiny.

For someone pondering the balance or imbalance of order and

chaos in the universe and the role or fate of humans in it, this assignment fit right in. I would be compiling not philosophical musings but a quantitative study of the place of order-loving bodies amid world-famous chaos. The study's premise was that humans are fragile and fleeting beings whose lives and dreams can be obliterated by randomness, by faulty decisions, by thirty seconds without oxygen, by a head hitting a blindly staring rock. Astronomers couldn't make such a study with their telescopes, nor poets with their metaphors, nor existentialists in their cafés. The National Park Service was actually paying me to conduct an official scientific existential experiment. I would not only be compiling statistics on how chaos treated humans but interviewing humans about their experiences and feelings, about the choices they make when facing chaos, starting with the choice to come here at all.

The NPS was also paying me, if unknowingly, to pursue an inquiry into the strangeness of ravens, into human perceptions of the universe and ourselves. And what, if any, was the connection between cosmic order or chaos and my vision of the raven? Did my vision arise from the order that had been searching for more order for billions of years and that was now announcing an even clearer truth? Or did it arise from the chaos that, here and elsewhere, tempted humans to risk their lives and thus deny the value of their lives? Did the raven's extreme presence have anything to say to the extreme absence that haunts humans and leaves them in existential despair?

Destiny was also calling from the sky. On the night the Colorado River flood would begin, a newly discovered comet would be making its closest approach to Earth, becoming the brightest comet in decades. It was in the northern sky and I wouldn't be able to see it from inside the canyon, but just knowing it was there would add to the epic, primordial feeling of the flood. And the moon would be full.

On my drive to Toroweap, the rim area overlooking Lava Falls, I stopped at Lees Ferry. Big satellite trucks from three national TV networks were installed on the boat ramp. The flood was big news. I talked with two river trips rigging to launch in the morning. The leader of a private trip was threatening to sue the NPS if anyone on

THE RAVEN

his trip got hurt. The leader of the other trip, organized by a leading kayaking school, was expecting the adventure of a lifetime, kayaking the entire river in flood.

With the national media not just reporting but overdramatizing the flood, a drowning would get major coverage and likely trigger major criticism of the flood and the National Park Service. One of the most likely places for a drowning was Lava Falls. I would need to document what happened, and why.

Toroweap is fifty miles off the highway, down a dirt road and a broad valley formed by a major geological fault, which near the canyon cracked open Earth's crust and squeezed up lava and built volcanoes, including the six-hundred-foot-tall Vulcans Throne above Lava Falls, and poured lava into the canyon.

On one visit to Toroweap I climbed Vulcans Throne, more cinders than solid rock, so with every crunching footstep upward I slid partway back down. At the top I did not find any Roman gods. I was hoping to find a vent but instead found a smooth dome. As the volcano sloped toward the canyon, it held a widening concavity. Looking south, across the canyon, I saw Prospect Canyon, clogged with lava, and beyond it Prospect Valley, a continuation of the geological fault down which I'd driven from the north. I realized I was standing directly atop the fault line. Looking at the contours of Vulcans Throne again, I realized it was cut in half, with one side slumping. In the eons since the volcano had died, the fault had continued sliding, carrying half the volcano with it. Suddenly the ground beneath me felt a lot less solid. I felt the enormous power pushing the continents around, pushing the ground into the sky, flicking aside the most powerful rivers.

The old contest between lava and the Colorado River hadn't ceased, for with every flash flood down Prospect Canyon tons of volcanic rubble poured into the river, rebuilding one of the largest debris fans in the Grand Canyon, rebuilding Lava Falls, refusing to let the river go free.

The first time I visited Toroweap I was startled by its view of Lava Falls, two miles downstream. Something that is overwhelming when

257

you stand beside it was just a tiny patch of white, barely fluttering, silent. Arriving rafters were just specks of color, even through my binoculars. All their tension and debates, then their wave-shocked screams and joy, made no impression up here. I supposed this was the way human life appeared to Vulcan or other gods.

Since Lava Falls is 194 miles downstream from Glen Canyon Dam, the flood waters would take a day and a half to get here, giving me a night, with very dark skies, to camp at Toroweap and enjoy Comet Hyakutake.

It was the brightest comet in decades partly because it was passing far closer to Earth than most comets. The comet's coma, its head, appeared larger than the moon, and its bluish-white tail covered a sixth of the sky, the longest comet tail most humans would see in a lifetime. Seen through binoculars the tail showed many streamers and pulses, the tons of material melting away every minute as the comet approached the sun. When this comet had last visited Earth, seventeen thousand years before, humans were stacking stones into the first towns, maybe into altars to ominous comet gods.

At its brightest, the comet tail stretched into the constellation Coma Berenices and was pointed at the constellation Corvus. The Raven. This was the sort of omen I agreed with.

The Babylonians created the constellation Corvus, a raven sacred to the god of rain, a raven perched on a serpent, the god of the underworld. The Greeks took over this constellation and added Apollo and a story about Apollo sending a raven to fetch water, but the raven delayed to eat and lied that a water snake had delayed him, for which Apollo turned ravens black and sentenced them to the sky forever, clutching Hydra the snake. To the ancient Jews, perhaps picking up on the Greek myth, Corvus was Noah's raven, supposed to bring Noah news of land, but he was disobedient and now forever trapped in flight. To the Tlingit, Corvus is a raven carrying the sun toward sunset.

I imagined that Corvus, with its godlike powers, saw the comet vividly. Corvus watched the comet appearing out of nowhere, revealing the universe's abundance, and plunging toward the sun and blossoming like a luminous spring flower. Corvus felt affinity with the

THE RAVEN

comet, for both understood the glory of soaring through space and shining through darkness. At the end of a 13.8-billion-year sentence, the comet was an exclamation mark!

I tried to see the comet through Raven's eyes, but I saw, first, transience and urgency. I knew that by the time I emerged from the canyon, the comet would be much farther away and dimmer. I would probably never again see a comet tail so long and brilliant. I tried to soak it up, to charge up memories that would last. Yet the comet was reminding me that human lives, too, are brief, unpredictable flares. I saw the comet dying back into the gas and dust out of which we both were born, saw a beauty inseparable from disintegration.

My effort to see the comet with maximum intensity seemed good practice for my project of seeing a Lava Falls raven with maximum intensity. A sky full of unusual intensity and a river full of unusual intensity seemed to be challenging me to reply, offering me their energy and beauty. If the intensity of this sky, this river, and this canyon could not induce a greater intensity in me, I was not sure what ever would.

Though I wasn't able to see the comet from inside the canyon, I would not forget that it was up there, and I would begin seeing it in displaced ways. Some of the water of the Colorado River had arrived on Earth as the ice of comets, helping to form the first oceans and rivers. As part of more recent comets, the Colorado River had stirred wonder and fear in humans, creating gods and telescopes. Grand Canyon waterfalls painted the images of ghost comets. Ghost comets felt the same gravity that had called them toward the sun and they plunged again, forming rapids.

In Lava Falls I saw quiet darkness erupting as a comet, roaring with energy, unfurling a long white tail full of streamers and pulses and wisps, its whiteness emphasized by the black cliffs, white even at night, gradually fading back into darkness downstream.

The descent to Lava Falls started on solid lava but soon involved steep slopes of cinders, which slid under my bootsteps. Then the trail stopped and I faced a long, narrow, very steep, dark ravine packed with loose talus so close to its angle of repose that a lizard's touch could set it sliding. I clung to the ravine walls as long as possible,

259

but the columnar jointed walls were fragile and tried to break off. Finally I had to go onto the talus, but you couldn't stand on it, not for long, so I had to sit on it and scoot down it, but I became a bulldozer shoving masses of rocks downward, some of which started rolling and bouncing, not stopping for a hundred feet, two hundred feet, some the size of bowling balls, the size of human skulls, rolling and bouncing and setting more rocks rolling and clattering loudly and echoing off the walls. My momentum was hard to control, even when digging my hands into the talus (I'd brought leather work gloves so my hands wouldn't get shredded) and braking with my feet. My heavy backpack added to my weight and prevented me from leaning back more fully. Sometimes all my braking couldn't stop me from sliding out of control for many feet. This was the hiking equivalent of running Lava Falls.

Soon after I arrived, the flood arrived. The water went from green to swirling brown tendrils to solid brown. Driftwood, even entire trees, that had been combed off the shores for 194 miles, paraded by. The river climbed up boulders and submerged them. The waves grew and changed shape. The roar grew much louder. A year ago a flash flood down Prospect Canyon dumped an unusually large debris fan into the river, and most of it was still here, but now it began collapsing, hundreds of boulders falling in and rolling down the riverbed with a bowling alley rumble.

I watched all this from the scouting ridge. I watched Lava Falls turning back into the monster it used to be. In an hour's time, hydraulic features morphed surprisingly. The ledge hole ceased being a hole and became a massive wave, regularly building up and crashing down with a thunderous whomp. The V-wave filled in its deep trough and became one of a series of huge waves. The bottom boulder disappeared and became a weird, darker wave. Waves grew bigger and wider. The whole rapid extended much farther downstream. The roar grew and grew. Logs and trees tried to run the falls and got tossed crazily, even snapped apart. I was awed, as if watching a volcano coming to life.

To get a closer look I scrambled down the shore of volcanic boulders. As usual, the waves got a lot bigger than they seemed from the scouting ridge. The water pouring over the bottom boulder became a ten-foot waterfall behind it, surging from moment to moment. The eddy behind the boulder was a madhouse of currents, surges, boils, undertows, and explosions. A swimmer falling into this would get slammed against the boulder and dragged underwater, but all this turbulence would probably soon flush them out, probably still alive.

I hadn't figured out where I was going to camp, and as I watched the river rise I realized that some of the cozy sand patches upstream from the rapid were going to be flooded or close to it. But right behind me at the scouting ridge, up a steep slope and tucked into the cliff, was a flat nook just wide enough and tall enough for a small tent. This cliff had been a lava cascade that had cooled from the outside inward and jointed itself into irregular columns, now overhanging, for the bottoms of the columns often break off. The slope above the scouting ridge was a mess of fallen columns, and so was the little flat spot. The overhanging columns here had conspicuous cracks between them. I reached in and touched a column and tried to wiggle it, like a kid testing the looseness of a tooth. I reassured myself that geological decay was very slow by human standards, and that if a rock did come loose in the night it would instantly hit my tent and be deflected sufficiently. This was an irresistible spot, the most awesome camping spot I'd ever have, with the scouting ridge as my front porch and a fantastic view of Lava Falls, all night in full moonlight. I would turn a place of fear into a wonder-full home.

As I grew enthralled by the rising river and changing rapid and parading logs, I forgot all about watching for ravens, until some birds nudged me. Far downstream a flock of ducks was skimming just above the water, heading for Lava Falls. Once I had seen a miscalculating duck fly into a rapid and get smacked down by a wave. When these ducks neared the falls they gained altitude, cleared the waves, and landed a hundred yards upstream. Ninety yards. Eighty yards. This was migration season, with visitors who didn't entirely understand the concept of powerful rivers. Forty yards. They began paddling upstream, paddling ever harder. Twenty yards, squawking in panic, they took off and fled.

Seeing No Evil

Perhaps it was surprising I did not perceive the Lava Falls raven as an omen.

Everyone perceives Lava Falls as ominous. Long before people hear it and see it, before they see the pleasant red cliffs disappearing behind funeral-black curtains, even before they begin their river trip, people hear the name Lava Falls being spoken ominously. On the day they will run it, rafters rig their rafts more seriously, with flipping a genuine possibility, with injuries easy to imagine, with deaths a historical fact.

Considering the dark symbolism human culture has loaded onto ravens for centuries, I might have perceived the Lava Falls raven as a warning of death. Ravens have been cast in the role of sorcery and evil and death in numerous mythologies, poems, fictions, paintings, and movies. Halloween imagery is full of ravens. For some peoples, the evil of ravens was dangerously real, requiring protective rituals. Given the real danger of Lava Falls, an ominous raven would have been appropriate.

Yet ravens also offer a case study in the superficiality and flexibility of human culture. It's only for some peoples that ravens are evil. For other peoples, ravens, or at least a mythological Raven, is a god: sacred, wise, magical, benevolent, a Promethean giver of light or water, even the creator of the world. The mythological Raven, like biological ravens, is a trickster, ingeniously outsmarting opponents. Raven sometimes offers serious moral lessons, sometimes plays a clown.

The dividing line between good ravens and evil ravens is largely geographical. Northern peoples, including Scandinavians, Siberians, and Native Americans of the Arctic and Pacific Northwest, have viewed ravens as good, while peoples of more temperate regions have viewed ravens as bad. These moral judgments derive from how ravens fit into human survival. For agricultural peoples, ravens and crows are raiders of crops. For northern peoples, who rely mainly on fishing and hunting, ravens are not competitors for food, and the ingenuity and fortitude with which ravens survive in a harsh climate is admirable, even magical.

Human events further scrambled the moral identities of ravens. In AD 793 Britain was Christian, but under the surface pagan beliefs and practices endured. Then the pagan Vikings invaded, with raven images on their sails and shields and pendants. Many British found raven worship a natural and appealing addition to their old pagan ways. For centuries to come, British Christians had to compete against Viking religion, and they did so by demonizing everything about it. Ravens were not wise and friendly tricksters but instruments of Satan, tricking you into Hell. This long anti-Viking campaign left Britain with an unusually demonic view of ravens, a view the British took to America and perpetuated in the images of Edgar Allan Poe and Hollywood horror movies.

Nor did I see the Lava Falls raven as a northern, shamanistic, positive raven, in spite of several things prompting me to do so. Kayaks were invented by the Inuit, who gave ravens a central and positive role in their cosmology. Many northern peoples made Raven the creator of rivers. In one story, all the world's fresh water was imprisoned and guarded by Petrel, but Raven tricked Petrel, filled his mouth with water and flew over the world spitting out water, forming streams and rivers. In another version, Raven didn't spit but urinated to create rivers. This story is less reverent than the stories of how the Ganges, Nile, and Yellow Rivers flowed out of heaven or the body of a god, but here too rivers were created by the same god who had created the order of the universe. Another prompter was a rock window overlooking Lava Falls, dubbed Odin's Eye. On the god Odin's shoulders perched two shamanistic ravens, Hugin and Munin, or "Thought" and "Memory," whom he sent out every morning to fly over the world and bring him all the news he needed to govern wisely. River lore holds that looking into Odin's Eye brings you luck in Lava Falls. But the Lava Falls raven was a desert raven who had never heard of the Inuit, and I had no time for Odin's Eye.

I did not see the Lava Falls raven with any tint of mythological roles or good or evil. Normally it's not easy to avoid all the meanings human minds impose on reality instantly and constantly, yet when you are seconds from dropping over Lava Falls in a kayak, human culture is very far away and worthless for defending you. Lava Falls blasts centuries of intellectual constructions out of you and leaves you naked before reality, a reality more powerful than ever.

This raven had also escaped from the zoo of ideas modern Westerners impose on nature, ideas transmitted through two centuries of Romanticism. When science began displacing the biblical cosmos, the Romantics replaced it with the testament of nature, which was filled with divine wisdom, creativity, benevolence, and harmony. To know God, you needed only to "commune with nature." British Romantic poets and American nature seers like Thoreau and Muir were not seeking evolutionary depths or ecological complexities; they were seeking God. Even as religious faith continued retreating and science continued adding more chaos to nature, people continued pursuing the old Romantic goals, often with now-empty postures: now "communing with nature" consisted of getting close to a wild animal, or enjoying natural beauty, or admiring ecological harmony even if it was drenched in competition and blood and extinction. In the Lava Falls raven I had not seen wilderness, the sublime, noble natural beauty or harmony, ecological roles, or civilization's victim or cure. All these images, too, were irrelevant and a waste of crucial time.

I had not seen a bird or an animal of any sort, not a life moving and individual, not a biological entity different from the geological and hydrological and astronomical entities around it, not a substance and shape and color defined by differences, but simply an entity, no time to story it or even name it, an entity stripped naked to the simple condition of being an entity, a presence, a pure and strong presence, an existence, and there was nothing further, nothing stronger it needed to be.

RAVEN MYTH: RAVEN SEARCHES FOR GOD

When Raven learned how fine-tuned the universe is, how every atomic pulse and water ripple and lightning bolt and tree and egg is lined up perfectly and pointing toward Raven, he decided that this couldn't have happened by chance and that the only rational conclusion was that the universe was carefully planned and implemented by a god. Ah, but which god? According to EGO, the Earth Gods Omnibus, Earth has 2,148,697 named gods. Or had. Just because silly humans had stopped believing in a god didn't prove it wasn't real, and indeed if it was a jealous and wrathful creator god who had been abandoned, this could explain a lot about the world today. Raven yearned to know the truth, to identify the one true god, to personally thank his creator. Raven began roaming the world and interviewing gods.

Raven soon discovered that when he asked gods if they were true gods or the one and only god, they always said Yes. Every creator god insisted that other gods claiming to be creator gods were imposters. Every river god and mountain god claimed all the credit for their realms.

Raven, being a trickster, was able to trick many gods out of their conceits. He asked Colorado River gods how long they had taken to carve the Grand Canyon, and if they said one year or one flood, he cited the geological facts to them. He asked sun gods about the temperature required to generate atomic fusion and light, but they were usually in the dark. He asked bison gods about the chemicals required to digest grass, and he eliminated them. He asked creator gods for the exact charge and spin and speed of electrons, and then told them they weren't up to speed. But many creator gods, especially the retirees, had been keeping up with the news and could recite scientific facts with high accuracy. Raven pondered how he could ferret out the one true creator god.

Every time gods claimed to be the one true creator god, Raven heaped praise on them and held up a photograph of a galaxy and asked them to autograph it and, better yet, impress their fingerprint on it, the way Hollywood celebrities leave their handprints in famous sidewalk cement. The gods were always flattered and pleased to comply. Raven found they were an egotistical bunch. When Raven analyzed their fingerprints, whether they were coyote gods or river gods or invisible gods or gods without any limbs, their fingerprints were always human fingerprints. Their signatures always said "human." Nothing but human. Now Raven saw why the gods were such an egotistical bunch, and why they were such sycophants to humans. The gods were just the excrement of the human ego, the same ego that makes such a mess of human lives and human society and covers the globe with warfare, the same ego that insisted the universe was made just for humans and has to be managed to meet their needs. Crybaby humans, neurotic humans, ridiculous humans—Raven never had much respect for the human ego and often enjoyed teasing it.

With his foolproof fingerprint method, Raven worked his way through the entire EGO address book of 2,148,697 gods and found that every one of them was a fake, and he wasn't shy about telling them so. Some gods readily agreed and reported that they had long ago gotten tired of being the puppets of childish humans. But other gods took themselves quite seriously and rejected Raven's message and refused to look at their human fingerprints and accused Raven of blasphemy and threatened to strike him with lightning bolts or send him to burn in Hell. But of course the gods didn't have any such power. Raven grew tired of this and became blunter, telling gods, "You are nothing but a human ego turd." Raven got into terrible shouting matches with gods, truly terrible.

When Raven had eliminated all 2,148,697 gods, he was left with a puzzle. Undeniably, logically, the universe had been designed by a god. From the character of the universe, Raven tried to deduce the character of the creator god. First, god was a genius, capable of designing such an ingenious universe. Second, god loved order, filling the universe with coherence and precision. Third, god loved chaos, filling the universe with the violence of stars and storms and deaths. Fourth, god loved to watch living creatures, whether out of love or for entertainment. Fifth, god didn't care about living creatures all that much. Sixth, god had a sense of humor, for he had let humans concoct 2,148,697 joke gods. Seventh, god had eyes, for god had filled the universe with light. Eighth, god loved blackness, for god had created so much of it. Ninth, god loved riddles, for god had left the universe a giant riddle.

Raven added up these clues. They all seemed familiar. They fit a pattern. Raven looked at himself: with his light-loving eyes he looked at his marvelous black body. He got out another photograph of a galaxy. He pressed his claw into it. He'd made only a claw mark, not a human fingerprint. Raven had a revelation. He was such a genius trickster, he had tricked even himself.

Raven was the one true creator god who had created the universe. This explained a lot!

I Become Lava Falls

For a week of drinking water I was going to rely on the river, but I hadn't thought about exactly where I was going to get my water. I walked upstream, but found that the river banks were too steep and unstable. If I fell into the cold, strong current, wearing boots and no life jacket, I could get swept into the rapid. I continued upstream to the eddy where rafts park, and found that its little beach had grown quite a bit bigger. The beach-building flood was already working! It was also a problem, for this eddy was doing a fine job of concentrating sediments. I filled a water bottle and held it up and saw that it was dim with silt, a tough job for my water filter.

I saw a solution, but it was problematic. Just downstream from the scouting ridge, just where the river plunged, was a large, turbulent eddy, probably the strongest eddy in the rapid, and this turbulence meant the eddy wasn't accumulating so much sediment. I climbed down from the scouting ridge and approached the eddy. The view from here was frightful. I was looking directly into the ledge hole, or rather the monster ocean surf wave it had become. To my right the river plunged out of sight. The eddy was surging back and forth against the boulders and shore, splashing the gravel well inland, splashing my boots and legs. But the eddy was pushing inward, not pulling a slipped body outward. My analytical mind reported that this was a manageable situation, but my emotional mind protested that this plan was crazy, terrifying. But it was safer than scrambling farther downshore on loose, wet, slippery talus.

I crept toward the eddy and started to dip a water bottle into it, but the water surged and filled the bottle in one gulp and nearly ripped it out of my hand. Note to myself: If you drop a bottle, *don't wade in* to retrieve it.

I held up the bottle to assess its turbidity. The water was still swirling from the force of the eddy. Many sand grains stood out from the cloud, drifting toward the bottom, starting to build a thin stratum. I was re-creating the sedimentation happening in the river, the sedimentation that had created the canyon rocks. The motion continued, if slowing, Lava Falls refusing to surrender its power. If I was to swallow this water right now I would be funneling the force of Lava Falls right into me.

Lava Falls would be my kitchen faucet for a whole week. Had anyone ever drank directly out of Lava Falls for a week? Not rafters. I wondered if Native Americans drank out of Lava Falls because Lava Falls was a god and they were on a vision quest.

At sunset I climbed back down to the eddy for more water. This became my daily ritual: two quarts at sunrise, two quarts at sunset. Overnight my bottled Lava Falls water became a lot clearer, and I filtered it into a third bottle. Toward sunset the ledge hole/wave, with light dancing on it, was especially eerie.

My act of filtering water acknowledged that I was not the river but an entity with a stricter hydrology, yet by drinking from Lava Falls for a week, Lava Falls was turning into me.

My water had been in the middle of shouting *Lava Falls!* when I interrupted it. This water had been jumping into the falls and feeling its directives and performing them. Now this lassoed dragon was calming down and paying attention and mastering new skills. Indiscriminate water that had formed the tongue of Lava Falls was now being tasted by my tongue and analyzed for its fitness, and soon it would become my tongue. Crazy currents were being swallowed into order, channeled into veins and cells, into highly regular and graceful waves, driven not by gravity and brute force but by a rising dance, the river's heart attack becoming a reliably pulsing heart, the river's roar becoming a voice that might speak more eloquently than Lava Falls.

RAVEN WATCHING, PART ONE

When I first arrived at Lava Falls and watched the river rising and the rapid growing in size and power and noise, and continual change in shapes, I was astonished and excited by it all, and I was grateful to be present to watch it. I watched it steadily for hours, studied all its changing details. Sleep could not wall out its power and novelty.

But within twenty-four hours I was growing accustomed to it. It was no longer new or changing. The waves were basically the same as they'd been the day before. The roar was the same. And it was less exciting. Lava Falls, even in flood, was becoming normal. The human mind was behaving normally, making the world normal.

On my first day I was so enthralled by the rapid that I noticed little else, not the cliffs or the sky or the wildlife. If a pterodactyl had flown overhead, I might not have noticed.

The next morning a raven darted before me and reminded me of my raven project. I watched the raven flapping and climbing and circling. I watched the raven. It looked like every raven. It was doing what ravens always do. It was an ordinary raven. Entirely ordinary. There was nothing special about it, nothing at all strange.

THE SOUND OF GRAVITY

Most of the canyon points upward. Each rock stratum grew from the bottom up, from life piling up fossils, rivers piling up silt, wind piling up sand. The strata piled up one atop another, a mile of upwardness, and then tectonic forces lifted the rock more than a mile upward. The canyon's spires and buttes and granite intrusions point upward.

Yet at Lava Falls the downward overcame the upward. From its volcano, lava flowed downward and down the cliffs and down the river for miles. The lava tried to stop and turn itself into rock but gravity commanded it to keep going. When it did become rock, gravity commanded it to crack apart and fall off and roll into the river. At Lava Falls, gravity ripped the river downward and the river fought back and jumped upward, but didn't get far. The same gravity that creates gravitational waves and black holes was creating these waves and water holes.

The lava had been evicted from its proper realm inside Earth and had tried to return to the darkness it still bore. The river was so determined to reach the center of Earth that it had tunneled a mile into the ground. The lava and the river obeyed the same force that keeps planets and galaxies swirling. The canyon was carved by the same force that had defied the dissipation of the Big Bang and summoned it into galaxies and turned darkness into starlight. Lava Falls was part of the same test that will determine whether the universe expands forever or slows and contracts. Lava Falls was here because of a force that had no obvious reason for being here or for having the strength and character it has. For no obvious reason, gravity had condensed creatures who stood here staring downward into the power and mystery of gravity, into the drama of a cosmos rising and falling.

In Lava Falls, gravity had folded itself into a voice, expressing the previously silent music of the spheres.

This roar was the voice of water everywhere, flowing through its cycles, through the ages, heard or not. It was the sound of rain gentle or storming. It was the sound of a million creeks and rivers and waterfalls. It was the sound of oceans relentlessly surging against the confining shore, oceans that in Lava Falls were screaming for home.

It was the roar of the ancient volcano above me, forcing water to flow through its lava-boulder teeth and impersonate the volcano's old power. It was the voice of mute creatures who had lived at the bottom of the sea and drank this water and become limestone, now boulders ruddering the waves. It was the sound of erosion, the ancient sound of the Grand Canyon growing deeper.

I hadn't thought about the roar. I would be bombarded by the roar of Lava Falls all day and all night. For boaters, this roar means danger and nervousness. Would it allow me to sleep? Even mild noises can disturb our sleep.

After the twilight zone becomes night, reality contracts and expands and bends. Reality loses most of the details humans need to move safely and find things. Even in your own home, moving around in the dark could mean stubbing toes and knocking things over. At my Lava Falls perch, stumbling was dangerous, and thus night was dangerous. But night also hides the details that distract and trick us into thinking we know the world, and it takes the false blue lid off the sky and reveals that our little human dramas are taking place on a greater stage.

Into this strangeness, consciousness melts, reality bends. Lava Falls did not keep me awake, but it did warp my dreams, almost into nightmares. When I awoke, especially if night remained, I lingered in dreamland for a while, but not trying to make sense of my dreams. I lingered in dreamland to hold on to its surrealism, its oddly bending streets and clouds, its judgment that the world is fundamentally strange. I wished I could apply this surrealism to the daytime world, but as morning spread and navigable details emerged and I thought of breakfast, the world became ordinary again. I grew accustomed to the roar and sometimes didn't even notice it. This was yet another measure, another omen, of how humans can barely notice the universe, even at its most extravagant, can barely notice our own presence in it.

RAVEN WATCHING, PART TWO

Clearly, I was not seeing the raven in the right way, with the right kind of attention. I needed to concentrate better.

A raven flew past, and I focused on it. I saw blackness, blackness like the lava around me. No, even blacker, for much of the lava was gray black and the raven was coal black. I saw a large, curved beak that said this was a raven and not a crow, but then there shouldn't be many crows around here, so this was a pointless point about a point. I estimated the raven's wingspan at three feet, maybe four. I watched its flight mechanics, more like a hawk than a crow. I estimated the raven was about two feet from beak to tail. I knew that its Latin name was *Corvus corax*. Did ravens in ancient Rome understand any Latin words? Were Corvettes named for *Corvus*? This might make sense, as ravens fly like sports cars. Did Corvettes ever come in the color black?

FALLING

One time when I was approaching my water-hole eddy, stepping onto the sloping, wet gravel, one of my feet slipped slightly, only three inches, only for a second, but this jolted me with alarm, though I was in no real danger of falling. When I'd filled my water bottles and stood up and started moving, I noticed that this water swayed back and forth, not in service to my motion and safety but in loyalty to the river's obsessive pursuit of gravity. You can take water out of the river, but you can't take the river out of water.

Except maybe by drinking it. As the days passed, an increasing portion of my body was composed of Lava Falls. Through my brain and eyes Lava Falls was seeing itself, but it was not recognizing itself. Lava Falls was seeing itself with fear. Lava Falls was one massive act of losing balance and falling, but Lava Falls had been abducted and brainwashed and now saw falling as a horror.

My climb between the eddy and my perch was also a bit problematic and made me ponder falling. Even on smooth soft carpets humans monitor falling, though we are gravity incarnate. Gravity did its best to stop the universe from expanding and evolving, to freeze matter at the Big Bang where all matter was the same, but it failed and the unity of creation was shattered and the universe became shrapnel of identity, hopelessly confused. As matter complicated itself into many forms, it began behaving as separate entities, sometimes conflicting entities—stars devouring comets, rivers erasing rock—and it forgot it was one uni-verse. When the universe masked itself with living faces, experiencing itself as biological needs and fears, it became even more convoluted, more of a stranger to itself.

So now Lava Falls was standing here, partly incarnate, finally with the power to see itself and say itself, yet it was seeing only an alien, a monster, and feeling fear.

With human powers of seeing and imagination, Lava Falls could see itself with X-ray vision from top to bottom, see thousands of currents falling and rising and swerving, combining and dividing and conflicting and dissipating, a vast gearworks of water, generating thousands of waves and splashes and sprays. With its new powers Lava Falls could see its old powers, the force it had been all this time. Lava Falls could gaze into time and see its ancient wizard's long white beard, see itself changing, see itself changing the canyon. It could see its volcano mother giving it birth, giving it enduring strength. And now, its outer force become inner force, Lava Falls would erupt with awe.

FEATHERS

When I spotted a raven flying over the river, I was ready to see a feather falling off and fluttering down. This must happen occasionally.

One time I found a raven feather lying on a Grand Canyon beach. I picked it up and turned it and admired its engineering and beauty. I stepped to the water, held the feather high, stem pointing outward, and tossed it over the river. It fluttered down and "landed" on

THE RAVEN

the water and floated away, spinning slowly, revealing the secrets of the currents.

This act was inspired by a conversation I'd had recently on a California river. I was sitting in an eddy when someone began admiring the craftsmanship of my wooden kayak paddle. He was the highest authority, for he had invented this shape of kayak paddle, the dihedral design. Tom Johnson was also a pioneering designer of kayaks and once the coach of the US Olympic kayaking team. Tom began pointing out the similarities of my paddle and a bird feather. Imagine, he asked, a bird feather dropping to the ground. It doesn't drop straight down. It flutters back and forth, and if it is falling stem first it spins around and around. This is because bird feathers are designed to maximize air resistance. A bird feather has a central spine that the feather curves away from on both sides, and the spine curves from end to end, making the whole feather concave, a net for air, making the air do some of the work of holding up a bird. When you combine feathers into a wing, with its further shapes, you get the best combination of stability, lift, and maneuverability. Tom had done a thorough study of the aerodynamics of bird wings and airplane wings and applied this to kayak paddles, which until then had been smooth, not resisting the water enough, wasting energy on forward strokes and diving too fast when you were trying to brace or roll. The same aerodynamics that makes a feather flutter and not dive to the ground is now—Tom pointed to my paddle's spine and curves—giving me extra leverage, better grab, when I am paddling and bracing and rolling.

I gazed at my paddle as if I'd never seen it before. I had seen only a superficial shape and beauty and not how I was interacting with nature's secrets. All along, I had been a bird, flying through a sky of water.

If a raven was watching me toss that raven feather on the river, what did it think?

On narrower rivers people scouting a rapid sometimes toss sticks on the water to help them figure out the trajectories of the currents and waves. People seldom try this for Grand Canyon rapids, which are too wide and long and turbulent, but I have seen it done at Lava

Falls, where the scouting ridge overlooks the confusing and critical right-side entrance. And if someone tossed a raven feather into Lava Falls as a probe, what would it report about the trajectories of order and chaos?

Thus I was waiting for a raven to drop a feather into the river, waiting for the erosion that had preened these cliffs into black rubble to reach into a raven. If I had found a raven feather on the ground, I would have tossed it into Lava Falls, and with the same spirit with which Hopis leave prayer feathers in the Sipapu.

Even as it falls through the air, a raven feather is still trying to fly, still proclaiming its ravenness. It touches the water gently and floats easily, then feels the winds of an approaching storm. Into Lava Falls the feather plunges and all the water turns white, but the feather alone denies it. All the water turns chaotic, twisting the feather, ripping at it, but the feather flexes easily and retains its form. Chaos pushes the feather down and farther down, but the feather tries to rise, rise to the sky where it belongs. Through chaos the feather-paddle tries to steer its way. The feather flutters like the battle flag of life. Lava Falls soon fades away, but the feather continues going, for miles, for days, merging with the river, helping the river wave its wings.

And in the night, the raven feather merges with the blackness of the universe. The feather was already the ambassador of the sky, and now the night sky flows down like lava and fills the river and the canyon, and whitewater becomes blackwater. The feather becomes the night's quill signing on the river the authorship of the Van Gogh swirls of Lava Falls. The feather floats past humans curled in their sleeping bags, and even though the night is signing its authorship of them for eight hours and signing in their dreams the fundamental strangeness of the world, they still might imagine themselves to be not only autonomous but the masters of the river and of themselves.

I never did see a raven feather run Lava Falls. Both ravens and the river were holding on to their secrets.

RAVEN WATCHING, PART THREE

Looking for ravens would have been easier amid white cliffs or red cliffs. Amid black cliffs the ravens were well hidden, even when they started moving. If camouflage is one of nature's priorities, this was the perfect place for ravens. Ravens seem oddly out of place in the Arctic, where most animals are white. Northern humans have various stories to account for ravens being black, usually with the trickster getting tricked.

I was looking for an anomaly, a shape or motion, against the black rock. I was looking for strangeness. All I saw was ordinary black rock. Entirely ordinary. The ravens were hiding amid the ordinary, pretending they too were ordinary. When night arrived and they melted into night, they pretended it was just another ordinary night.

CROW MOTHER

I once commissioned a Hopi friend to carve for me a Crow Mother katsina doll, whose most distinctive feature is black wings emerging from both sides of her head and swirling upward. The Hopis don't make much distinction between crows and ravens, at least for religious purposes. As farmers in a food-scarce environment, Hopis should view crows and ravens as enemies, maybe as symbols of evil. Yet for Hopis, Angwusnasomtaqa, or Crow Mother, is a highly positive spirit. She is the mother of the other katsinas and leads their return from their home on a distant mountain to begin their half-year ceremonial cycle in the Hopi villages. After a February night of ceremonies in a kiva, she emerges at sunrise carrying bean sprouts, an overture of the good harvests to come.

My friend was from Third Mesa, the mesa closest to the Grand Canyon and the most likely destination of the Hopi ancestors who had migrated out of the canyon. Looking at his hands, I saw thousand-year-old hands that had lifted and stacked Grand Canyon rocks into houses and kivas, coaxed Grand Canyon soil into corn, and shaped Grand Canyon clay and pigments into pottery. His hands, full of erosional lines, would now pick up a cottonwood root and erode it into a crow/raven face that would guard the world against erosion. My Crow Mother katsina would stare at me with the deep strata secrets that for a thousand years had drawn humans back to the canyon to meet the powers of creation.

One February my friend invited me to his village for the year's first katsina ceremony. I remained on the kiva roof, in the dark, where differences between humans disappeared and I was freed from the outsider status I would have had in daylight; more than once I was mistaken for a Hopi. But the night had not hidden our true identities; it had revealed them. We were mysterious shapes, contours of the night itself. We were a shapelessness that with thousands of ceremonies all over Earth cried out to the steadier and interpreting stars.

From a mile high in the desert, the stars were abundant and vivid. The stars looked over our shoulders as we midwifed a meaning for them also. A few faces close to the rooftop entrance glowed ghost-like from the lantern light within. I heard chanting, rattling gourds, and especially drumming, the steady, strong, hypnotic, heartbeat, echoing drumming.

I was hearing an echo from the Grand Canyon. This kiva held a symbolic sipapu. I thought of how at this moment the canyon's Sipapu dome was silently watching the sky, like an observatory dome, the starlight sparkling on its water, on ripples of emergence.

Somewhere around me, in plazas or gardens or these bodies, was Grand Canyon soil that had been carried out of the canyon not by the river but inside the humans leaving it forever. Somewhere in nearby fields grew the descendants of corn that had grown in Grand Canyon fields, corn still blooming in these bodies and rattling in their gourds, kernels jumping randomly yet calling out against randomness. The shadows on these kiva walls had grown from giant campfire shadows on Grand Canyon cliffs.

I would hear this drumming for days afterward, and nights. I heard it the next time I entered the canyon and stood beside a remnant kiva. I heard it when I saw finger impressions in potsherds or granary mortar. I heard it in thunderstorms delivering the blessings of rain. I heard it in the drumming of woodpeckers and in the silent motions of snakes. I heard it in the drumming of Lava Falls and in my own heartbeat when I awakened before sunrise and waited patiently for light, as if I might see, emerging from the black rocks, a supernatural Crow Mother giving her blessings to life.

RAVEN WATCHING, PART FOUR

I was keeping a constant watch for raft trips appearing around the upstream bend. Every two minutes or so I glanced upstream, which was unnecessary, for the upstream bend was far away and rafters would take a good while to arrive here and much longer to scout the rapid. I was too eager for "something to happen." Nothing was happening with the cliffs and plants: they were the same from hour to hour. But surely I couldn't feel that nothing was happening with Lava Falls, with its incessant and massive action. Then again, its overall appearance and roar wasn't changing much, only small details. I was getting so habituated to Lava Falls that I was turning away from it in hopes of seeing something new, some unpredictable human dramas. Even Lava Falls could not override the habituation functions of the human mind.

The same thing was happening with ravens. I found that a raven sitting still was less interesting than a raven in flight. To where was it flying? Would it be acrobatic? Would it reveal a favorite cliff hiding place? Flap, flap, flap, action, action, fascination. A do-nothing raven was nothing.

RIVER WATCHING, PART ONE

The first rafters to arrive were a private trip with five rafts. They started scouting carefully, but soon one rafter declared he was ready to go, declared "I can go anywhere I want in that!" He walked back to his raft, a small raft, untied it, and pushed off. No one was riding with him, no one to high-side or bail or rescue a swimmer. Running a major rapid without other boats in the water violates a basic safety rule.

The first wave ripped both oars out of his hands. Out of control, his raft hit a giant wave at the wrong angle and was slam-dunked over. The upside-down raft careened through the rest of the waves. Finally he surfaced and got to the raft and held on, but he couldn't pull himself onto it. He was still clinging to it when he disappeared around the distant bend. He was still clinging to it three miles later when a camp of geologists saw him, blue in the face, no doubt hypothermic, with no one pursuing him. They rushed to help, but with chattering teeth he started cursing them, yelling to leave him alone: he could handle this by himself. He still had not conceded that the Colorado River was stronger than him. He was less concerned about his life than about his image as a forceful man. The coldness of space was soaking into him.

When the geologists got him to shore and righted his raft, he got out a bottle of whiskey and drank half of it.

THE ZUNI RAVEN

I crossed paths with a ghost.

I was heading into the canyon, heading for Chimik'yana'kya Dey'a, the Zuni name for Ribbon Falls, where humans emerged from a series of underworlds into this world. I met the tracks of something emerging, something mysterious. One day 315 million years ago a reptile had gone for a walk and left its footprints in some sandy mud. What had it seen and smelled? Of its whole life, of maybe a million footsteps, only twenty-eight footsteps had been recorded. I was looking for these footprints because a few days before it was announced that these were the oldest reptile footprints ever discovered, anywhere in the world.

I'd heard this news on a radio station while camping in Death Valley, after a day when I walked all the way across the mostly dry salt flat, leaving my footprints in a moister section, footprints that might last for years. I'd seen dead dragonflies now heavily encrusted with salt, and in the middle of the salt flat I'd found the remains of a raven, torn apart by some raptor, its skull and bones and feathers sparkling with salt, at night sparkling with moonlight and the Milky Way.

The canyon's reptile tracks were exposed by a recent rockslide. The Zunis say that when humans first emerged into this world, cosmic order and human forms were not entirely fixed, and some people, in crossing the river, reverted into amphibians or reptiles. I sat at the reptile tracks awhile, resting my ancient and unstable feet.

To Ribbon Falls I was carrying a Zuni raven fetish. It was two inches tall, carved from black marble, with grooves to indicate wing feathers and tail feathers, with specks of turquoise for eyes.

THE RAVEN

Geologically speaking, black marble is not really marble but limestone loaded with bitumen, the stuff of coal, which meant that this raven held the swamps and dragonfly sips and dinosaur footprints of a million Cretaceous days and nights, truly held the powers of trickster transformations that humans would ascribe to ravens. For Zunis, fetishes carry and transfer supernatural power from animal spirits to humans, to help bring rain, growth, or healing. In the Zuni creation story the Zunis wandered away from Ribbon Falls and the Grand Canyon in search of the Middle Place, the place where they belonged. They split into several groups, each choosing a bird egg to guide them, and the group that chose the raven egg was the group that found the right place.

The next morning, from my camp at Phantom Ranch, I headed up the trail a few miles to Ribbon Falls. Zunis make pilgrimages to Ribbon Falls, gathering water and plants for rituals back home. I seemed to be taking my Zuni raven on some sort of a pilgrimage. But what kind?

Ribbon Falls glowed with a thousand sparkles as water flowed down it. Where it was covered with moss, it sparkled green. It was a mineral dome, perhaps two dozen feet tall, with a flowing shape because it had condensed out of the minerals flowing down it, flowing from a waterfall from the cliff above. Amid miles of dry cliffs, such a waterfall felt surrealistic. Who needs the Egyptian pyramids when you've got a dome as magical as this?

At the dome's base the water flowed into a clear pool, then became a little creek that soon joined Bright Angel Creek and flowed into the Colorado River, meaning that for the rest of its journey through the canyon, the river was sacred emergence water.

I took the raven fetish out of my backpack and set him in the gravel beside the pool, looking at it. His turquoise eyes looked happy about the water—for many southwestern tribes, turquoise symbolizes water, precious water. His smooth black marble was shiny in the sunlight. Okay, I said (silently), you should know more about this than I do: what's the correct ritual for a raven fetish at Ribbon Falls? But soon some other hikers arrived, and it turned out my raven was rather shy, for he jumped into my jacket pocket and hid there until they left.

Feeling safer, the raven climbed out of my pocket and waded into the water, only his head protruding. *You could consider this a baptism*, said the raven, *and not necessarily a Christian baptism. There are flocks of religions that use water as a sacrament, that define immersion as a blessing, a purification, a healing, a fertilization, a connection with origins, a pathway to the afterlife. Flocks of peoples turned waterfalls into the homes of spirits. But how many peoples made a waterfall their place of emergence or creation? You are seeing the soul of a desert people, for whom water is creative and sacred. Consider me touched by the creative and sacred.*

I carried the raven up the slope behind the falls and set him down atop the dome, in the falling water, and left him there a long time. The raven had been polished by the hands of a Zuni carver and now it was being polished by the water hands that had carved and polished those Zuni hands. Now the raven would hold erosion and shininess from the source of form and life. Erosion was theologically correct, for the Zunis hadn't imagined this as their Garden of Eden, monotheistically perfect. They had emerged from a series of imperfect underworlds and would have to journey through many further imperfections.

I looked down at the dome and at the water splashing onto it, both eroding it and building it higher. I saw geological forces become beauty and music. I saw psychological forces, for Ribbon Falls was also an idea, an erupting volcano of longing and meaning. And through this black marble the earth was reaching out and grasping for greater form.

That night I set the raven outside my tent, under the Milky Way and the moon, to merge with the night, its eyes more blue stars. I wondered if a ringtail or a pack rat would come and inspect it, even raven-nap it, but it was still there to greet the dawn.

I tucked the raven into my backpack and headed out of the canyon, on a trail that for a mile followed the river—for the Zunis, Bright Angel Creek and the Colorado River were umbilical cords that connected them with the outside world.

I caught up with a group of a dozen hikers whose clothing contrasted with that of all the hikers wearing the trendiest outdoors fashions: they were obviously Amish, with the same color scheme as ravens. I had seen them the evening before, gathered in a circle near Bright Angel Creek. Now I asked them if theirs was a religious pilgrimage and they said not really, but certainly it was easy to feel God's presence and power here.

When I got to the reptile tracks, where bright serpent feet had trod, I wondered where this reptile had been going. It was not going on a symbolic journey, not searching for its emergence place or God. It was not carrying a raven fetish. Yet perhaps, already, this reptile had recognized the sun to be its god, the source of its warmth and motion. Already these tracks were on their way to becoming my own, including my tracks in the salt of Death Valley, where I had seen a raven as death when I could have seen it, amid miles of nothing but rock and minerals, as a brave emergence.

Of course, I realized: I was performing the proper ritual now. Raven was emerging from Ribbon Falls and the Grand Canyon, emerging into the emerged world.

RAVEN WATCHING, PART FIVE

A nudge happened, grew into a trend, then into hunger, so I dug out my bagels and peanut butter and, yes, they smelled good and looked good and tasted good to my biological self, and soon—*swoosh*—to another biological self, for a raven landed on the cliff nearby and began studying me and my food.

The raven was likely puzzled, for he seldom saw humans eating here, above the rapid. Many of the rafters scouting here were already starting to feel queasy, and they didn't want to add fuel to this unease. Ravens often saw humans pull into the beach half a mile below the rapid and have a feast, and sometimes humans stayed there overnight, so ravens watched for food opportunities. But I had to remind this raven of its national park contract: in exchange for no hunting or bulldozers, you get no handouts. The raven objected: *being wild, genuine ravens means getting food any way we can, finding it, stealing it, begging for it, killing it.* I didn't feel like debating the ethics of ravens, which might not be much different from the ethics of humans. I thought of the stories I'd heard of ravens stealing food from river camps, waiting until the chef turned away, swooping down and carrying off a pork chop. Some thefts required not just outsmarting opponents but mechanical ingenuity. Ravens have watched hikers zip food into a backpack pocket, waited until the backpack was unguarded, swooped in and grabbed the zipper and tugged it open, and pulled out the food and fled. On the rim one raven has pushed open the metal flap on a trash can and let its trusting partner hop inside and rummage for food. At least I and this raven agreed, as we stared at each other across our different shapes, that we were thoroughly biological creatures, cursed with hunger, happy at eating.

Food for thought. In my eagerness to see a raven as a portal to the strangeness of reality, I should not deny the hardness of reality, of creatures sentenced to survive through any means possible, condemned to steal one another's lives to maintain their own, forced to inflict pain and sorrow. The greatest numinosity could not hide this darkness.

BIRD SINGING

I was watching a raven perched on a boulder on my side of the river, when she took off and flew over the river and disappeared into the black cliffs. There was another sense in which she had disappeared. This side of the canyon was Grand Canyon National Park, but the opposite cliffs were Hualapai tribal lands. By crossing the river, the raven had gone from being a national park raven and a natural raven to being a Native American raven, part of a spiritual cosmos in which she holds magic powers of creation and guidance.

Years later I attended one of the central ceremonies of the Hualapai tribe, called bird singing. Along with hundreds of Hualapai, it drew members of other Yuman language tribes, who had traveled hundreds of miles out of the Mojave Desert. The Hualapai had once lived in the Mojave Desert, where they were masters of survival in one of the toughest environments on Earth, until the same drought that brought down Ancestral Puebloan society forced them to head for higher ground, the Colorado Plateau, which offered more coolness, moisture, plants, and game animals. Some of the Hualapai discovered Havasu Canyon, a branch of the Grand Canyon, where reliable

THE RAVEN

water and good soil made farming possible, and they changed their lifestyle enough that they considered themselves a different people, the Havasupai, many of whom were here tonight. The Hualapai and Havasupai and many Mojave Desert tribes still had much in common: language, dress, basket styles, music, foods, and spirituality.

The bird singing went on all day, a hot summer day with people hiding in the shade of a long, curving ramada roofed with juniper branches, built just for this ceremony. After dark, until midnight, people spread out under the stars. A long succession of people sang, shook gourds, played drums and flutes, and danced. The dances portrayed the motions of birds, with quick, short little steps and little hops, the dancers turning one-quarter or one-half around. The songs are based on bird calls. Some Yuman tribes live along the Colorado River, on the California/Arizona border, where lots of birds winter, and the Yumans welcome the annual migrations and know these birds well. Each tribe has its own spirit bird, which they express in unique songs. The Hualapai's bird spirit is the bluebird. For the Mojave tribe, it's the raven.

Bird singing and dancing is a long tradition, with about five hundred songs. Some songs are sacred, used in funerals or in all-night ceremonies in which carefully sequenced songs depict Yuman creation stories, tribal migration histories, and life lessons.

Since I could not understand the Yuman language, the meaning of the songs was lost on me, and I was not going to ask. But I'd heard there were sunset songs; so as evening arrived I supposed I was hearing birds welcoming the night. I watched twilight fading, the mountains disappearing, the stars emerging. The birds tapped, hopped, and spun. The birds sang, with occasional sharper cries of alertness.

These bluebirds and ravens and other birds had flown out of bird bodies and into mammalian bodies, so big, so heavy, limbs so bloated and leaden and dysfunctional. They continued trying to leap back into flight but couldn't get anywhere. They felt blinded by these short-sighted eyes yet also saw things with more clarity. They flew out of the ornithological world and into the ontological world, the world of mystery and meaning. They tapped out spiritual cryptography, shamanistic codes to the limbo in which life had found itself stranded and amazed.

Above this small flock doing their best to find patterns, the stars rolled with their own patterns and their own disorder, and within the birds the stars flowed, the night became ravens fearful and questing, merging into and emerging out of greater identities.

RAVEN WATCHING, PART SIX

Sometimes I didn't see ravens, only heard them calling. For other birds, humans would call it "singing," but raven voices do not fit human tastes in music. I tried to follow the calls to the callers, looking back and forth and up and down, but often I couldn't locate them. Then I wondered why I needed to actually see the ravens. Wasn't this only the prejudice of a vision-centric animal? Other animals rely on other senses. For canyon bats, sound is the guiding element. Coyotes make sense from scents. Snakes can map out heat sources and ground vibrations. And who are humans to be proud of our vision when owls and falcons have far better eyes than ours? But human eyes can feed deeper visions.

The next time I heard a raven calling, I tried not to look for it. I closed my eyes and listened. Humans may be vision-centric but we do love sounds, so much that we have filled the globe with music. If raven calls don't qualify as music, I was not so tempted to enjoy it as music. I just heard it as a sound, not silence, a presence, not nothing. I tried not to "visualize" the raven making these sounds but to let the sounds remain pure presence. It may be another trait of our being vision-centric that humans register invisibility as mysterious. Many of our gods are supposed to remain invisible. The phrase "the unseen" can mean ghosts and all the rest of the paranormal. So I left the raven unseen to coax more mystery out of it. I heard a cosmos full of silence, deep and never-broken silence, planets and moons of violent silence, and I heard there a sound, tiny but symphonic, a cosmos announcing its presence.

STAYING BALANCED

Be careful, a strange voice inside me was saying. Be careful, as I climbed down to the river and leaned over the surging water to draw my water. What voice was this? The be careful voice was more insistent than it might be in town, where help for accidents isn't far away. You are valuable, said the caring voice. This far from town, in this society of rocks and river, the voice obviously wasn't referring to my value in human society. You are valuable, said a voice we usually mistake to be our own. But this voice was going strong when I was born. Whose voice was this? It was the same voice speaking in every coyote and lizard and bird around me. It was an ancient voice that had spoken, in many thousands of accents, in the life now fossilized in the cliffs. It was only because trilobites and fish had heard this voice that we are here today, still hearing it, hereing it. When did this voice begin? What does it know? What authority does it have to tell me and lizards that our lives are good and should continue?

This voice seemed to be talking inside all the trees and flowers and cacti around me, for they too were answering it. In them the voice must be working in a different way, speaking silently, being heard unconsciously, offering further evidence that this voice is mindless, not requiring minds to operate. Trees and flowers don't need feelings or philosophies or moralities to convince them that living is important.

The same ordering drive that showed up in life also told water to rise into clouds and form rain and birth rivers and reincarnate the ocean. It convinced a nebula to turn into a reliable star and into Earth. The same voice that warned me to be careful about my gravity and balance had instructed galaxies in gravity and balance and motion and shape, including my matter in previous shapes. My atoms had billions of years of experience at holding extremely precise balances between particles and forces and other atoms. Some voice had told them, in atomic language: *It's extremely important that you do this.* In some form my balance, the order foundationing my life, was already present in the Big Bang, powerfully active and careful, and flew out of it toward the day it would whisper: *Be careful.*

The Big Bang would have been strange enough if it generated only light, but it was far stranger for its "urge" toward order, which one day would become the conscious urges of animals. Elementary physical forces had precisely the qualities required to turn the Big Bang into our universe, to provide a grand stairway up which matter

CANYON AND COSMOS

could climb into greater complexity and ability. If parameter after parameter had been even slightly different, this process would have shut down at step after step, leaving a universe of nothing but energy, nothing but hydrogen atoms, nothing but nebulae, nothing but cratered planets, nothing but volcanoes and rivers.

Early on, humans noticed that the world was a good fit for them, providing the sunlight, weather, water, soil, plants, animals, and babies they needed to live, a world puzzlingly imperfect but basically generous. It was reasonable to imagine that this gift had a giver, so humans imagined gods, who declared *be careful,* declared the value of life, more decisively. Science has refined the world's fitness into far more details and decimal points and implausibility, so it becomes even more reasonable to ask if this extremely precise design indicates a designer. But the universe's sophistication doesn't match any of the amateur creator gods humans have imagined.

The nature of matter not only allows for order, it forces it to emerge. Yet this necessity also produces contingency, massive amounts of it. Physics may require electrons to bond to atomic nuclei, but it doesn't care exactly which electrons find exactly which nuclei. Gravity may mandate that nebulae gather and condense into stars, but there's lots of contingency involved in sorting atoms into different nebulae and stars, and in deciding the numbers and sizes of stars. Supernovae could launch atoms toward future planets or toward an eternity of drifting through space. Necessity may require a star to have planets, but contingency decides if they consist of rock or water or gas or ice. Chemistry may be inclined to generate life, but contingency rules whether it happens or not and how life evolves. For the universe as a whole, life may be inevitable, imprinted on the cosmic dice, but massive amounts of chance go into every species, and every individual life is wildly improbable. From the Big Bang onward, at every step of cosmic evolution, necessity and contingency cooperated and conflicted, married and gave birth to children blending both of them.

Drawing my water, I looked up at the massive wave in front of me. Necessity demanded that water had to fall here, an entire planet demanded it, but contingency was deciding which water molecules went which way. Necessity was giving the rapid basic shapes, but contingency was deciding how waves jumped and intersected and changed from second to second, and which water went into this eddy. When I drank Lava Falls, necessity and contingency poured into me, blending with the necessity and contingency already here. The Big Bang was here, so insistent, still going strong. Old stars were here, the stars that sent my atoms toward Earth. Ancient rivers were here, rivers through which necessity and contingency had propelled me toward this meeting of waters. I could feel necessity and contingency swirling and mixing within me. Necessity had tried to create me but had gotten lost. Contingency had tried millions of times to block my creation but had gotten tired. They negotiated and agreed to let me live. Necessity and contingency got along because they were born from the same mystery, though necessity was the mystery of cosmic origins and contingency was the mystery of cosmic outcomes. I stood there feeling incalculably unlikely, downright impossible, infinitely strange.

ALL THE SAME

News flash! Peripheral vision to brain: motion! Define it, immediately! Don't forget when humans were only mice in danger of being stepped on by dinosaurs or snatched from the air.

My mouse brain instantly decided: bird, size, location, altitude, direction, speed, color, shape, species, intention, no threat, maybe entertaining. My human brain added a name and stood by with concepts like behavioral patterns and mythological roles. From no level did my brain offer: *existence*. This would serve no purpose. Existence was obvious. The obvious requires no observation or reaction. Something's simple existence isn't going to hurt you or feed you. It's something's *thingness* that needs to be defined, its differences from other things, its specific character, its relationship to you, its value for your own thingness. From mice analyzing the differences in dinosaur footsteps to Carl Linnaeus classifying the differences in mice and reptiles, brains are obsessed with differences. This smothers our ability to recognize that the first truth about things is that they exist, and that the state of existing does not hold any differences.

Everything in the universe is the same. On the plane of existence, size and form and function make no difference: everything exists equally. A pebble exists as much as a cliff. A tree exists in the same way as a planet. A raindrop and a river flow from the same source. A star and a firefly glow with the same power. Ravens flying away from me or toward me are equally here. Ravens are another manifestation of myself.

In my exploration of the identities of humans in the new cosmos, I started by mapping out our differences and connections with the cosmos, with atoms and stars and life. But now it's time to forget all of this. These connections are superficial. Our deepest connection with the cosmos, all along, is not between things but in being the same thing, the same reality, the same existence, all along.

I turned and watched existence flying away from me, flying toward being here.

River Watching, Part Two

When river trips arrived, they worked changes in my attention. Because my personal goal here was to maximize my attention, I was paying attention to the workings of attention. For hours I watched the rapid, nature's power independent of the human world, but arriving rafts shifted my mind back into seeing Lava Falls as an adversary with obscure routes and dangerous tricks. I snapped back into my role as an official observer with a checklist of variables I was supposed to record, such as how long people scouted, requiring me to watch not slowly migrating shadows but quickly advancing numbers on my watch. One day the first raft trip pulled up on the other side of the river, which meant I watched them from an impersonal distance. When they had run the rapid I had to write a report on how they'd done. All this stirred up mental sediment, which took awhile to settle down and allow me to see the rapid more clearly, as simply itself. Then a raft trip pulled up on my side of the river and people hiked up to me and I stood up and greeted them, and I was thrust back into the hall of mirrors of human relationships. I saw the river through their eyes as they worried, assessing and debating routes and consequences. I had to ask them about their experiences upstream. They seemed pleased that the National Park Service was officially registering their heroism. Quickly these strangers became friends, and I worried about them as they ran the rapid. When they disappeared around the bend, I felt a bit lonely. Again it took me awhile to recover from my divorce from the river.

TURBIDITY

I was at Lava Falls as part of an experiment in turbidity. To be precise, some 850,000 tons of sediment was being mobilized from the riverbed and creating beaches up to three feet tall. It occurred to me that I was conducting a further experiment in turbidity, the turbidity of the human mind, how the human mind fluctuates from awe to boredom.

Lava Falls fills everyone with awe, awe in its original definition as a mixture of fear and wonder. Lava Falls in flood invaded my senses and filled me with awe. The question was: am I going to become accustomed to Lava Falls? When it came to getting a good night's sleep amid this jet roar, I sure hoped so. But what about my awe? Was it, too, going to wear off?

The waves and the roar continued, hour after hour, day after day, basically the same, but in my mind they were diminishing. Drinking from the heart of Lava Falls started as a wild idea but was becoming a routine. Ravens were refusing to behave like anomalies and were becoming normal, with predictable habits. My own presence here was becoming normal, maybe in no danger of becoming just another day at the office, but no longer astonishing me when I woke up. And I was sleeping well enough. I was also sleeping when I was awake. My alertness was fading away, day by day, a dementia spreading on its own schedule. Even the mighty Lava Falls in flood could not escape from the human mind's habit of turning the entire universe into mere habits.

I was the victim of an evolutionary curse. Life had made a soul-selling bargain with a devil. Life had decided for its creatures: if you want to live, you will have to give up living. Or some of it. Biological creatures need to be adapted to their environments, and for animals this includes being mentally adapted. Survival takes a lot of energy, learning takes a lot of time and attention and mistakes. We have to classify every sight and sound as useful or not useful or a problem, and dismiss everything irrelevant to survival and let it fade into a reliable background. We can't afford to waste mental and emotional energy relearning the same places and things and lessons every day.

I gazed at some small rocks around me on my perch. Had I noticed them before? Of course I had, for I had moved them aside to make a smooth and safe little habitat. I had defined these rocks entirely by their fit with human purposes, not their geological stories or aesthetic values, and now they were irrelevant and required no further attention.

The same dementia at work on these little rocks was also applying itself to the entire canyon. I had inspected and classified the roar, the rapid, the cliffs, the plants, the lizards, and the ravens and filed them away as familiar, nonthreatening, requiring no emergency alerts. I seemed to have little control over this process; it was happening to me automatically; but this process, too, was so familiar that I hardly noticed it. My "don't notice" programming was also programmed to cover itself. Humans have always been this way, probably all animals too, with even fun serving a function, even surprise serving survival. And thus humans step outside their familiar and boring walls and walk beneath the starry sky and look up and are not surprised, not at all surprised by an entire universe. Or our own bodies.

Yet there is one trigger for which the human mind is programmed to respond with surprise and alertness. Change. An unfamiliar thing or event.

If I awoke one morning and found that my discarded and ignored little rocks had arranged themselves into the words "LOOK AT US," I would indeed look at them, for a long time. If the ravens had turned bright yellow, I would be shocked. Even if a change was entirely natural, it would fire up my alertness. If an opposite cliff suddenly calved off, I would jump to my feet and watch the boulders crashing. If a bighorn sheep came along, I would ignore everything else and watch it.

Human insensitivity to the familiar and arousal by new stimuli distorts our perceptions and behavior in many ways. People equate "living" with activity. People who haven't glanced at the sky for years will get excited by a comet or eclipse. People bored with every day will

THE RAVEN

celebrate wildly at the coming of a new year of every days. Women and men madly in love become too familiar, boring.

Most animals have a limited amount of novelty in their environments, and a limited ability to generate their own novelties. But humans have turned their novelty arousal response into a world of novelties. We rush from new to new, from style to style in clothing, music, food, art, movies, and speech. We are enthralled by sports and gambling because every action is unpredictable. We leave home for exotic places. We go rafting because every bend in the river is new. We love stories, in words and books and movies and plays, because every *Once upon a time* promises a long series of new events. We pursue science because it promises new stories.

Our religious creation stories indulged in the greatest novelties humans ever created, fantastic supernatural events. Yet even creation stories could not override the habituation functioning of the human mind and left the world explained, entirely normal magic. When humans see rocks and ravens and stars, their classification process does not even include the category of "existence." The human mind grants this automatically and jumps to assessing a thing's identity, location, motion, and potential. Often, humans glimpse existence or nonexistence through muddled biological eyes, confusing them with mere life and death.

I did notice the river's turbidity decreasing as its sediments settled out, maybe not from hour to hour, but definitely from dawn to dusk. I saw it in the river's color and in the opaqueness of my water bottles. I wanted my mind to follow the river's example. I imagined my mind's emotional mud and intellectual sand settling out, its clarity growing. Maybe this was working: I seemed to be seeing the rocks more clearly. The next time a raven flew by, I seemed to see it more vividly, as itself, whatever itself was.

RAVEN MYTH: RAVEN EATS THE GODS

When Raven was searching for the one true creator god he not only got into truly terrible shouting matches with fake gods but once became so exasperated that he beak-stabbed the god, killing it, and since Raven was hungry he went ahead and swallowed it. It was a blueberry god, the best tasting blueberry Raven had ever eaten. Raven became curious and began eating other gods. Ant gods, elephant gods, tree gods, river gods. All of them tasted great, some richer than others, Earth goddesses richer than imperial gods. He continued eating gods, nearly all of them.

Raven was also tasting another quality, elusive to define at first. It was some sort of numinosity. It glowed with value, beyond nutritional value. Raven finally realized he was tasting sacredness. It was not the sacredness humans assigned to gods, which varied a lot depending on their usefulness to humans. This sacredness was already in the mist, the mystery, from which humans congealed gods.

As Raven ate more and more gods he felt his stomach and mind filling with gods, felt the gods blending together, their boundaries dissolving, their contradictions canceling out, leaving essential godliness, more and more godliness, greater and richer consciousness. From different gods he felt different forms of consciousness: creation mind and destruction mind, good mind and evil mind, law mind and trickster mind, female mind and male mind, ritual mind and mystic mind, all combining into one supreme mind. Consciousness flowed through him like Lava Falls. He saw everything, every rock and tree and star, glowing with creation, with impossibility, with value not assigned by humans, with the sacredness that had created gods. He saw the gods as a continuation of the universe unfolding its order through fourteen billion years, now unfolding it as gods who judge that creation is good and encourage it onward. Through Raven flowed all the forces of creation, focused and peaking, returning to their source, to the mystery only further hidden by the word "sacred," and the universe saw itself with the greatest intensity ever.

Depths

I was getting frustrated by my inability to see beyond forms and colors and motions and ideas and to see simply existence, pure and strong. If the human mind is not well endowed for this, what could I do to prompt myself to see better? I was staring, vaguely, at the cliffs when it occurred to me that the canyon had been offering me a useful metaphor.

We don't notice existence just as we don't notice the ground on which we live. We take the ground for granite. We may know vaguely that it holds great depths of rocks, but we live on the surface, superficially, caring only about our smooth floors and lawns. Yet when people come to the Grand Canyon they have a revelation energized by the big contrast between the existence of rock and the nonexistence of rock. It's only because of a great absence that we experience a great presence. The rock layers we see in the canyon are exactly the same a few miles away, under forests and parking lots, but no one stands there and gawks at the idea of those strata beneath. Only nonexistence makes those strata real, gives them light and color and shapes and emotional charge. Every canyon contour, every ridge and butte and mesa, maps out the boundary of existence and nonexistence. The Tower of Ra and Buddha Temple exist only as acts of resistance against nonexistence. The sunrises and sunsets are beautiful only because they flow into nonexistence and find and paint existence. Ravens fly through nonexistence to continue their existence.

I looked at the cliffs again, at the contrast between rock and no-rock, color and invisibility, shape and formlessness. The contrast grew stronger. Nonexistence said that I usually don't notice it because it doesn't exist. Existence said that I usually don't notice it because it is all that exists. But now, together, nonexistence and existence were contrasting and emphasizing each other. The cliffs were making a statement about presence, and when I touched a cliff, possible only because I was moving in absence, I was touching existence. But only its face, with hidden depths beneath.

The river, too, is full of depths, but hidden. The river is full of images, images of cliffs, trees, clouds, and occasionally people, and Lava Falls is a riot of images, but images only on the surface. Descending, the images soon blur and disappear. The river is deep. Deeper than the superficial images in which humans dwell.

To see the river more strongly, to see the canyon and ravens and humans more deeply, we need to compare them with nothing. Absolutely nothing. We need to imagine a universe that never existed, with no Big Bang, no galaxies, no lava, no ravens. We need to imagine space without one speck of light or matter anywhere, complete

CANYON AND COSMOS

emptiness, not one twitch of change for eons, and even this may be too generous, for space with nothing in it remains a place, every place, and a universe that never existed should be no place, with no time. We need to imagine nothing, absolute nothing, not emptiness but nothing, not nothing forever but nothing ever, a nothing that could never imagine itself and leaves us unable to imagine it. Nothing but nothing.

Nonexistence is the only thing that should have existed. There is no justification for our universe of light and atoms and rivers to exist. There is no logic to it, no demand, no inevitability. It should not be here. Nothing should have remained nothing. Nothing could search through itself and turn itself inside out forever and never find anything. Nothing could add to itself and subtract from itself and multiply itself forever and remain nothing. Nothing could never think of any reason for something to exist.

Neither could humans. As logical animals, humans readily recognized the illogic of a universe that exists. This bothered us because human minds demanded to figure out cause and effect, but we could find no ultimate cause for the universe, or for the universe's flaws.

We tried to evade the illogical universe by creating creation stories, but they too were infected with illogic. In many stories, rivers or animals were already there, and the only question was how rivers or bears or ravens got to be exactly the way they are, or how humans arose from the animal world. In more ambitious religions the universe was allowed to emerge from nothing, nothing merely physical, as long as it emerged from the mind of a god. But as soon as logical humans asked the next obvious questions, asked where a god came from and why god acts the way god does and what this eternal god was doing for the eternity before creation, theologians told them to stop thinking and shut up. When human logic took the form of science, it was hugely successful at tracing effect to cause to effect to cause, at tracing one set of events, forces, particles, and laws back to previous sets of events, forces, particles, and laws. But even if science pushes the creation of the universe back to one ultimate event, force, particle, or law, science will face its own version of the question of where god came from, for there will be no answer as to why that first event, force, particle, or law exists and why it has precisely the qualities it needed to create a universe. Scientists, reluctant to admit that logic ultimately gets swallowed by an illogical universe, that all their equations from particle accelerators and telescopes add up to nothing, can become just as dishonest and turtles-all-the-way-down as theologians trying to hide from mystery.

If we can begin to imagine total nonexistence and the logic for it, we can set up a contrast that can prompt us to see objects more deeply, see them as, first of all, simply existence. Of all the qualities that define an object, the most obvious and important should be simply that it exists. Before we jump to exterior qualities like size and shape and color, before we start on identities like animal or mineral, before we go into internal realities like atoms, we should begin by seeing everything as a bold violation of nonexistence.

But human culture has neglected to explore this terrain, and our language for it is clumsy. Philosophers talk of "being" but quickly bury it under a mass of intellectual abstractions, sometimes trying to hide from it. Ultimately, existence may be beyond the grasp of images or language. However we try to imagine and speak of it, we should emphasize existence as outrageously illogical, with no ultimate basis, an invasion of the nothing that should have been everything, a reality castle floating in a surrealistic sky.

Why don't humans perceive existence as their first reality and dwell within it more fully? Our simple existence could be the core of human identity and a leading source of human value. But no, the human mind skips right over it. Our brain programming is preoccupied with how physical things behave and relate to one another. We perceive existence and nonexistence through merely biological eyes, as merely life and death, and death is threatening. If our minds have trouble imagining complete nonexistence, it's hard for us to hold the contrast that would emphasize existence. Because existence is so prolific, creating endless new forms, we easily get lost among them and forget that we and all forms are variations of the same existence. Whether religion is hiding from or celebrating existence, it often ends up smothering the mystery of it.

Yet here, deep in the earth, immersed in the forces of creation,

CANYON AND COSMOS

isolated from human society and its superficial images, more elemental realities and surrealities were trying to emerge and testify.

A raven flew by, but maybe it wasn't really a raven. Many northern peoples have ceremonies in which ravens appear, but they aren't really ravens, they are simply appearances, simply masks, sometimes very elaborate masks, behind which are humans who maybe aren't really humans but solidifications of cosmic mystery, through which cosmic mystery, too powerful to be suppressed even by dull minds, is breaking out, a question reaching for an answer but finding that all human questing only arrives back at the question.

A raven flew by, only the mask of a question.

THE WHITE RAVEN

White buffalo are rare, yet they are born often enough that Native Americans developed mythologies about them. I had never heard of a white raven, except in the world of mythology. But a reliable friend reported that a white raven had been seen on the canyon rim. Researching further, I found that white ravens do happen but are very rare, a one-in-a-genetic-million event.

This seemed a mythic event too. In the lore of many northern peoples, Raven started out white but got changed to black. He was stealing light or water for humans but got stuck in a chimney or got too close to the sun and got singed black, a mark of his nobility. Once again there was a geographical disparity. In temperate climates Raven's blackness was usually a form of punishment, resulting from curses by dignitaries like Apollo, Noah, Jesus, Mohammed, or God.

I had to go look for the white raven.

I started by checking in at places that are reliable hubs for the latest news: the park library, the barbershop, the mule barn. No one had heard about any white raven.

I headed for the ravens' favorite playground, the campground. I found a picnic table where seven or eight ravens—none white—had torn open a plastic garbage bag and spread its contents into a buffet. I walked all the loops of the campground and saw many ravens, all black. The next morning I walked the campground again, then walked miles of the rim. Many ravens, none white.

296

THE RAVEN

That night I went to the park's annual star party, with two dozen telescopes set up near the rim. One telescope belonged to a canyon old-timer, and when I asked him if he'd seen a white raven, he pointed to a constellation just above the horizon: Corvus.

I spent the next day wandering the village area. No white raven, no news of one. That evening I went to the ranger campfire program, given by Stew, who sometimes gave a raven-themed program titled "Nevermore." Stew had already installed, next to his wife's tombstone in the park cemetery, his own black tombstone with the sole word "Nevermore." Stew wondered if the appearance of a white raven was a mystical event, equal to a white buffalo or a Moby-Dick.

Again, Corvus presided over the star party. At dusk the real ravens had headed home, where they merged with the night. Somewhere the white raven, glowing faintly from the light of Corvus, stood out. I wondered why, when most of the universe is black, some peoples considered black a form of punishment. The northern storytellers who turned raven from white to black recognized the advantages of being black in an arctic climate. While other birds were forced to flee winter and undertake long, dangerous migrations to warmer climates, ravens absorb maximum sunlight, allowing them to stay home all winter. This was a trade-off, making ravens more visible to predators, but this disadvantage was offset by raven cleverness.

If judged by northern cosmologies, the canyon's white raven seemed a throwback to an earlier stage of creation, a karmic reprobate, perhaps sent to a place of creation to complete some mission.

The next morning, the summer solstice, I was driving into the park and passed something black lying beside the road. I pulled over and walked back. It was a raven. From a tree above, a raven squawked at me, likely the mate, mourning and upset. The dead raven bore a wound, and its eyes were gone. Beetles? Mice? Other birds? Ravens are usually smart enough to avoid being hit by cars: this dead raven was a violation of the natural order.

With the mate upset at my presence, I started to leave, but I felt it wasn't right to leave the raven to an unnatural end. Should I bury it? That, too, would be unnatural. This was probably not a Christian raven expecting resurrection. Maybe I should bury it in a natural tomb, in the canyon, letting it take one final flight through the sky it had loved.

I touched the raven feathers, their ruffled silk. When I tried to lift the raven, it was thoroughly limp, its head and tail flopping. I was surprised and distressed. A life had surrendered its shape, its will, its identity, its wings strong enough for storms. I looked into the raven's non-eyes and saw no one there.

Even a dead raven was too real to see as any symbol of death.

As I placed the raven in a bag, its mate scolded me. I told it: you can come along and watch.

I parked and walked a mile through the forest to the rim, carrying the bag in my hand, a little funeral procession. At the rim I looked for the steepest cliffs. I should have taken the raven out of the bag and given it a graceful toss, but I only let it slide out of the bag, which deflected its flight. The raven fell only five feet and landed on a tiny ledge barely big enough to hold it. I was dismayed. The trickster lives. Soon the raven rolled off and fell, wings actually fluttering, and disappeared. But it had flown only a few hundred feet and landed on a wider ledge, right on the edge, its tail feathers and wing extending over the edge, flapping in the breeze, flapping as always, flapping against death.

Within an hour, a squawking swept past me. Two ravens swirled downward and landed on the ledge, beside the dead raven. They were very upset, hopping about and talking fast and loud, perhaps talking to the dead raven. I should have considered that this place was the territory of other ravens, who knew it and its residents well. Now a complete stranger raven had shown up in a strange way, inexplicably dead. The two ravens reached out and tugged on the dead raven, as if testing it for life, for an explanation. A peregrine falcon victim? They would not guess that a human had been trying to apologize for a car accident, trying to pay tribute. The ravens tugged on the dead raven, pulling it away from the edge, inches away, a foot away, until it was no longer hanging over the edge and in danger of falling. One wing was left even more exposed to the wind and continued flying.

The ravens flew away, not for long. They returned to the ledge several times but kept their distance from the dead raven, landing on the very edge, almost falling off, flapping to stay on. They stared at the dead raven and squawked.

Not giving up on the white raven, I drove east to Desert View and stopped at all the overlooks along the way. No white raven. I returned to the raven grave and watched the afternoon shadows unfold from the cliffs like wings and grow blacker, finally embracing the dead raven. The resident ravens, still baffled by the mysteries of life and death, perched on a dead tree overlooking the cliffs and, near darkness, flew away.

I never found the white raven, or heard further news of it. It had come and gone like a ghost. If I were more mystical I might have decided that it was indeed a ghost, the pre-ghost of the dead raven, come to summon a sympathetic being to give it a grand burial, to tug it away from a graceless fall.

RAVEN MYTH: RAVEN MEETS ETERNITY

When Raven discovered he was the one true creator god, he decided he was entitled to the powers he had seen fake creator gods possessing. Omnipotence. Omnipresence. Omniscience.

Raven's greatly expanded awareness astonished and overjoyed him. Now he could see the entire universe, every planet in every detail, every rainstorm and rainbow, every mountain and pebble, every lizard and sparrow. It was dazzling. Then some Christian pointed out that as part of his legal contract as an omni-god, Raven was required to inhabit every human and take care of every falling sparrow. Raven focused on the minds of humans, and he found a terrible cacophony. He watched thousands of sparrows falling every hour. *Wait a minute*, Raven replied to the Christian, in some translations of the Bible, God only has to *know* about sparrows falling, not *care* about them. The Christian answered: A mere technicality, but you are obligated to care about humans, if you want to go on enjoying omni-powers.

As soon as humans realized that Raven had created the world, they began pestering him with questions and complaints and demands. They wanted to know why the world wasn't perfect, why it forced them to work hard and have accidents and arguments and get sick and get old and die. Raven answered with total honesty that

CANYON AND COSMOS

he didn't know. He didn't know why he had always been imperfect or why the world turned out imperfect. And there was nothing he could do about it.

Humans didn't want to hear this. They wanted Raven to be perfect. They told Raven their theories. They claimed that Raven was perfect but that there existed a demonic power who opposed Raven and marred the world. They claimed that Raven's creation was perfect but that humans had messed it up. They claimed that human lives deserved to be imperfect because they'd committed imperfections in previous incarnations. Theologians and philosophers came to Raven with elaborate rationalizations about "the problem of evil," and villagers came to Raven crying about sore toenails or gambling loses. People tried everything to escape from imperfection: sacrificing sheep, magic spells, praying, meditating, grand ceremonies. They pleaded to Raven relentlessly.

Raven tried his best. He monitored every human impulse, need, emotion, neurosis, dumb thought, fall, fight, and fart. It was overwhelming. He grew impatient, and when humans demanded to go to heaven, Raven replied that Christians never provided a heaven for ravens so why the hell should Raven send humans to heaven? But it was in his contract. Raven was being driven crazy. He decided he needed a psychiatrist.

He went to the best expert on these things, the Christian god. Raven started by complaining about Luke 12:24 in which in his Sermon on the Mount Jesus says: "Consider the ravens, for they neither sow nor reap . . . God feedeth them." What a con, declared Raven; I had to work my tail off to feed myself. And then Matthew 6:26 changes "consider the ravens" to "consider the birds," as if Jesus is ashamed to associate with ravens.

Better be careful, replied Dr. God. He cited Psalm 30:17: "The eye that mocks God will be plucked out by the ravens." "You might," said God, "have to pluck out your own eye. At least now you understand what I've had to put up with for centuries. Here's a paradox to consider: One of the imperfections of an imperfect creation is that there is no one to answer the question of why creation is imperfect. When imperfect creatures try to answer it, they only create further imperfections. The only perfect thing in the universe is its mystery. In that, creation has reached absolute, inscrutable perfection."

Dr. God added that if Raven found humans exhausting, just wait, it would be worth it. Raven didn't understand, but he waited, through eons. He did notice that every time a planet of human types died out, a little space in his awareness was less distracted and he saw the universe more clearly, not just its forms but its existence. As whole galaxies went dark, raven's awareness grew steadily brighter.

When the final living planet died, God announced that it was time for him to go. His contract was for presiding over life, nothing more. But you, Raven, were never bound by that contract, and you can live forever, with omni-powers. With that, God vanished.

Raven endured. Raven watched the galaxies going dark. Raven didn't mind blackness. Raven watched the black galaxies receding into the blackness. Black galaxies were just as existent as bright galaxies. Dead galaxies were just as existent as living galaxies. With no further distractions, Raven luxuriated in existence, exalted in existence. Raven would have eternity to love his creation. Eternity, eternity, Raven's heaven.

EXALTATION MYTH: THE CRUCIFIXION

When God grew tired of an eternity of loneliness and decided it would be nice to have someone to talk to, decided to create the universe, he considered creating a pantheistic universe in which everything is a manifestation of one divine being, but this would mean he was only talking to himself, which he already had been doing for eternity, so he decided to create a material universe in which all bodies, nonliving or living, would be separate from one another, giving him genuine others to talk to, and others who would look upon him not as my own divine body but as their creator, requiring greater gratitude. But even at the moment God let there be light, he realized that his long neurotic loneliness might be infecting his creation, filling it with loneliness, with creatures pathologically separate from one another and from him. Soon enough, God was being bombarded with complaints about loneliness and the problems caused by separateness—indifference, egotism, cruelty, crime, war—but he was also hearing about how separateness led to happiness, including love and marriage and gratitude for God's friendship.

Still, God felt far outside his creation, still lonely. He wondered what it felt like to be a living body with a heartbeat, walking upon the solid earth. He decided he would incarnate himself in human form. He would not only satisfy his curiosity and try out this love thing but clean up some theological roadblocks and make it possible for humans to rise to heaven and become angels and give God better company. Always proud of creating trees and knowing from the start he would end up crucified on a tree, God decided to be born to a carpenter so he could immerse himself in wood.

God experienced birth more intensely than any human ever had. God was shocked, both thrilled and horrified, by the feel of a human body, of matter itself, of having shape and solidity and motions and locations and boundaries beyond which there is no *you*. He cried. God was frustrated by human senses, which so muffled and distorted the world, but nevertheless God was able to experience the world far more intensely than humans could. Starlight hit him with the force of a supernova. Walking, he felt the size and strength of the entire planet. The lightest breeze was a symphony of both sounds and scents. God had always seen time as infinite, but now he lived it, felt it moving, every new second a miracle. In a forest, God saw life's every shape and sound and scent and action not as his blueprint for creation but as amazingly real.

Working with wood, God was enthralled by the upward force of trees, the embodied time, the scents released by sawing, the artistry of woodgrain, the strength upholding houses, the fire waiting within. God caressed the wood and loved its reality and the hidden shapes he was liberating. God had created the world in such a rush that he hadn't taken time to enjoy his own craftsmanship, and now he reveled in the smallest details.

For thirty years God kept his identity and mission a secret to give himself more time to enjoy being human. But one day he felt compelled to perform a healing, and everyone was astonished and called it a miracle. *Silly humans. Everything is a miracle, every fish and loaf and sunrise, and doesn't stop being miraculous just because it's recurring.*

It was time. God's plan for the crucifixion was to intensify humanness so much that it would transmute the leaden human soul into gold worthy of heaven. God knew it was his own fault for making humans so dull, and now he would make amends.

Dragging his cross, God felt wood more deeply than ever, the sheer reality of it, the holiness of it. Being nailed to his cross, God felt bonded to wood more intensely, felt his own body more intensely than ever. His pain was a rocket carrying him into the unexplored depths of his body. His intensity was a revelation, an ecstasy. Howl, howl, hallelujah. He could have spent his time on Earth as a tree, never feeling anything. He gazed at the crowd, at his mother, donkeys, Roman soldiers, believers and doubters, and he marveled at the mystery of bodies coming out of other bodies, bodies of different shapes, bodies loving other bodies, bodies so cruel to other bodies, bodies trapped in mystery. God felt life yearning for life. God wavered between life and death but wavered on an outgoing tide. God stood in the doorway of existence and felt the vacuum of nonexistence pulling him out, felt the full contrast between existence and nonexistence, felt its ultimate intensity, an intensity he'd not felt since the opening door of the Big Bang. God exclaimed: *Holy! Holy!*

God saw a raven flying straight toward him, saw her wings so vividly, looked right into her eyes and saw himself there.

Swoosh, went himself.

The raven landed on the cross, right next to God's sideways head. Hungry and not waiting for death, she pecked at God's hand wound, the blood still trickling out. God found this new experience fascinating, but it did offend his pride. God scolded: "Where were you when I laid the foundations of the earth and laced the sweet bands of Orion? Who provideth for raven his food? Who do you think you are?"

Raven answered: "Quark, quark, quark."

"And how did you know you consist of quarks? It will take humans a long time and huge particle accelerators to figure out that matter is made of quarks." God realized that for all their smallness, quarks contained room for ravens to hide, and for all its brightness, the Big Bang contained shadows for ravens to hide and wait.

The raven bent over God's face and went after the tenderest, juiciest part, the eye. Blood flowed down God's cheek and onto his lips and tasted like a great river.

The raven now saw with God's eye, saw wood and faces and the earth and the sky with infinite intensity. After a lifetime of perching in and grasping trees, the raven was finally grasped by a tree, she fully inhabited a tree, felt the intensity of creation within it, and she exclaimed: *Holy! Holy!*

Now when the raven flew, she could fly at the speed of God, much faster than the speed of light, so fast, absorbing visions so quickly, that it overwhelmed her eyes and accelerated her brain, turning it into a particle accelerator reaching deeper into matter than human physics ever could. She was generating a shock wave in space-time, a crack in reality. As she generated the leading edge of the shock wave, reality could not keep up with her and left a trailing edge of pure Absence, letting her constantly feel the vast difference between Presence and Absence, being and nonbeing, letting her see existence with the greatest intensity. She saw the stars glowing not with light but with being. In galaxies and rainbows, she saw not just aesthetic beauty but the beauty of being. In animals and plants, she saw not just biological lives but the life of being. In adding up the galaxies, she saw the massiveness of being. In everything, she saw not just matter but how being mattered. As being flooded through her, it accelerated her into ecstasy.

FLOWINGS

As I continued diverting water out of Lava Falls and into myself, as the river flowed into me and set my mind flowing, flowing, flowing, I began to see flowings in things around me.

I saw the black cliffs not as rock but as a river, a flowing of magma up from the earth and into the air and over the land and into the canyon, wave after wave of lava flowing into the river and embracing the river as another form of itself. I saw the black cliffs still flowing today, grains and rocks and boulders falling into the river and joining its celebration of flowing.

Lava had turned into soil and flowed upward into plants and sent roots into the river-dampened soil and even into the rapid itself, so

Lava Falls was flowing upward into little green waves that grasped sunlight and married river and sunlight into a new flow.

I began to see the sky flowing, which was easy enough from the clouds flowing across it, and then the clouds became apparitions of the invisible, of flowing moistures and pressures and temperatures and gravities, of the planet turning and the sun burning. The sky's blueness flowed out of vast prisms and atomic secrets. The sunset colors were the taillights of the moving Earth. Shadows, the spies of night, crept out of hiding and took over the canyon and handed it to the cosmos, brilliant with flowings.

As I watched ravens flowing over the river, I saw that they too were the river and lava flowing. These ravens had been drinking out of the river all their lives; so unlike me, they were a hundred percent river. The ravens were eating lava berries, lava seeds, and lava bugs. I saw motion, motion strong, motion ceaseless, motion merging with the winds, motion become voices, motion with destinations, motion in one of the millions of shapes in which motion incarnates itself before moving on. The river had risen into raven blood, raven wingbeats, heartbeat waves, breath waves, DNA waves, brain waves. The river was defying the gravity of Lava Falls and flying with willpower, emotion, and awareness. The river was ravening. The lava had erupted again, been resurrected into black angel wings flowing across the black cliffs.

In the night, in a cosmos brilliant with blackness that had flowed into the forms of lava and ravens, the ravens sat on the cliffs above me, watching the river and the night, merged into the lava and the night from which they had been born, out of which they flew every dawn as if in ritual reenactment of their creation.

EXALTATION MYTH: THE TAO

The stream flowed so steadily, so primordially, so beautifully, so calmly even where turbulent, that a few months from now a man would decide that a stream was the best metaphor for cosmic energy and wise human living, and he would call it Tao.

The stream flowed naturally, without trying, without any goal, because its nature was to flow. If it were not flowing, it would not be itself. It did not cling to any shape and was stronger for being shapeless. If it ran into rocks, it did not waste time worrying about them but parted and went around them and reformed itself downstream. It was the most yielding of all things, yet eventually rocks, the firmest of all things, would yield to it. It had no inside or outside but was constantly turning itself inside out. It didn't try to cling to shore, to stop motion and change, for then it would destroy itself, and in not trying to stop time, it became timeless. It did not cling to any sense of self, and thus it could go from being rain to snow to ice to springs to streams to rivers to waterfalls to lakes to oceans to clouds. If it clung to its identity as a stream, it could never become life. It quenched the thirst of trees and deer but not out of seeking gratitude. It was beautiful but could not admire itself without leaving itself. It was colorless but magnified all the colors around it. Its invisibility made deeper things visible. When it was still it was voiceless, but when moving it allowed even rocks to speak. It did not give itself a name or speak of itself: it was content to remain mysterious. In some form it had flowed from the beginning of the cosmos, but it did not know anything about its origins. Even a bear with its mighty claws could not grasp it, only be filled and nourished by it. It constantly drew yin/yang symbols, even shaping the stones into them, but it never worshipped them. It was always falling toward the abyss but never ended up there and was never fallen. It never worried about ceasing to flow, and when it did stop being a stream, it was content to be the ocean.

Then the stream was diverted from its perfect Tao and into a human body. It filled the body as it would a bottle, taking on its shape, including its thoughts and feelings. The water was appalled. Things were so chaotic in here. Even the order in here was chaotic. Everything was separated into little boxes, with walls everywhere that wouldn't allow water to merge and flow and be itself. The entire package was a larger separation, a body that imagined itself to be an independent entity, fundamentally different from water or other life-forms. This package gave itself a name to define itself as different: Lao Tzu. Lao Tzu's mind polluted the water with rioting desires and fantasies.

THE RAVEN

Lao Tzu was having the most frustrating day in a very bad month. To smother his harsh feelings he stopped by a café and got drunk. He collapsed onto the floor. As he lay there in a dreamy haze he overheard the water within him talking to itself about what a mess he was, what a waste of energy. He listened, carefully. He heard his water—no, not *his* water—saying that humans would save themselves much grief if only they could flow like water.

Lao Tzu decided to try out water's way. He began to flow like water. When he ran into an obstacle, he parted and went around it and reunited, without worrying that he was breaking up his identity. For months he went on trying—no, not trying; water doesn't try. He went on learning. He had been this water all along but not recognized it, had forgotten how to flow.

People came to Lao Tzu to learn from him, but he knew he was still living his ideas—no, not his ideas—on the surface. So one day he rode away on a water buffalo—it could not have been a mere land animal. He found a steep, narrow Himalayan valley with a stream flowing through it, flowing into cascades and crystal-clear pools. He gazed into these pools to give himself greater clarity. He saw thousands of rounded and colorful rocks that had yielded to the clarity of water and taken on its identity. He saw fish for whom the entire universe was water. And one day, finally, he fully became water.

He called out to his fellow waters to come and see. And the waters came, from near and far, streams and waterfalls and lakes and rivers. Even the Ganges came to Lao Tzu to be enlightened. The Nile and Amazon and Mississippi, hearing the rumor that rivers need not be obscured by mud, came to see. The Colorado River, too, came to clarify things. Lao Tzu, as real and not metaphorical water, told them that his teachings to humans had been flawed. Confused and pitiful humans may need to be like water to find more peace. But truly, the highest state of being is not being Tao water but being water in human shape, with human senses and consciousness. It is true, continued Lao Tzu, that humans are rapids, disturbances of the peace, obscure at reflecting the sky or the trees or themselves. They may need to pretend to be Tao or immortal souls or reincarnated tigers to find peace. But turning water into wine and walking on water are cheap tricks compared with water turning into life. Baptism in the Jordan or bathing in the Ganges are superficial compared with water becoming living bodies. Oceans full of water-stirring gods are not as profound as oceans becoming gods. It might be trouble to be human, but it also gives water eyes and voices and touch and passions and awareness and the greatest intensity. It releases water from being only water forever. It lets us recognize ourselves as genuine Tao, cosmic energy flowing from mystery. Let us celebrate our brief time in human bodies, our gift of life, our time as water gods.

RIVER WATCHING, PART THREE

When at Lees Ferry I'd met the dozen kayakers on the kayaking school trip, they were dreaming of conquering the Colorado River, but most had never been here before at any river level and they were nervous about it.

By the time they got to Lava Falls, four kayakers had taken such a beating that they had given up and were riding on the rafts. Two others scouted Lava Falls and decided not to try it. The six who tried Lava Falls did okay, four of them right side up. They conquered the lava fifty thousand times older than themselves, conquered the river they never glimpsed within themselves.

GOETHE

Living beneath Vulcans Throne, looking up at the lava cascades above and all around me, living inside a little nook of that lava, waiting for lava rocks to fall on me, it was easy to feel the awe humans have always felt for volcanoes, for the mysteries of the earth, for the conflict between creation and destruction. This awe once inspired Goethe to climb Mount Vesuvius.

Beneath Vesuvius's rolling black clouds, scrambling up its massive black slopes, Goethe was a tiny speck being hit by a steady drizzle of ash. "At a sufficient distance," he would write, "it appeared a grand and elevating spectacle." But as Goethe climbed, Vesuvius became more ominous. The ash was hot, some of it even glowing. His companion, an artist, refused to go farther and took shelter under an overhanging ledge. "To him," Goethe wrote, "the artist of form, who concerns himself with none but the most beautiful of human and animal shapes . . . such a frightful and shapeless conglomeration of matter, which, moreover, is continually preying upon itself, and proclaiming war against every idea of the beautiful, must have appeared utterly abominable."

Goethe stuffed his hat with silk handkerchiefs and continued the ascent. "The presence of danger generally exercises on man a kind of attraction, and calls forth a spirit of opposition in the human breast to defy it." Goethe was feeling his inner Faust. He was also seeking the Romantic sublime, nature's revelation of its power and divinity, against which tiny humans should tremble in awe and worship. Goethe was sure Vesuvius had something to tell him, not in any scientific way, although Goethe was betting his life on his observation that Vesuvius was spouting lava at regular intervals, on his calculation he could get to the top and back down before the next spout. Goethe wanted to measure Vesuvius through the gauge of human feelings, letting Vesuvius reveal its depths of power and beauty and horror. He wanted to stand on the thin boundary between, on one side, the beautiful ocean and the "light-hearted and merry life" of Naples, and, so close, the abyss.

Far below Goethe, tiny in the distance, the lava-buried ruins of Pompeii and Herculaneum held thousands of bodies or cavities shaped like bodies. The day after the citizens of Pompeii and Herculaneum had observed the annual feast to Vulcan, tossing into ceremonial fires small fish, symbolic of human lives, Vesuvius exploded.

To the early Romans, Vulcan was a god of destruction, his temples located safely far away from cities. Later, under the influence of the Greeks, Vulcan absorbed the characteristics of the Greek god of fire and blacksmithing, Hephaestus, who was primarily a creative god. Vulcan became a rare thing among blacksmith gods, both creative and destructive. With his hammer and anvil, Vulcan forged the golden shield of Achilles and the lightning bolts with which Jupiter upheld the order of the world. When Roman society was mainly agricultural, Jupiter's lightning bolts brought rain and a plentiful harvest, and as Rome became more urban, Jupiter's lightning bolts became the enforcer of justice among humans. Vulcan's forge was located inside Mount Etna, whose fire and smoke and thunder meant that Vulcan was working. Romans living at a time when Mount Etna was merely smoking viewed Vulcan as a force of creation, while Romans who saw Mount Etna explode viewed Vulcan as chaos.

At last, what had been a "grand and elevating spectacle" from afar became a black horizon in front of Goethe, and then he stood on top, four thousand feet above the sea. He gazed into the volcano and saw—nothing. Only smoke, emerging from "a thousand crannies" and rolling upward. He waited, and for a moment the breeze thinned the smoke enough for him to glimpse rock walls broken by cracks, but soon they vanished behind the veil again. He stood and waited, like Empedocles.

Goethe's contemporary, the German poet Friedrich Hölderlin, wrote a drama, *The Death of Empedocles*, about the ancient Greek philosopher who had thrown himself into Mount Etna. At least, Hölderlin assumed this is what happened. History knows only that Empedocles vanished, leaving one of his sandals atop Mount Etna.

Empedocles combined previous speculations into a cosmology in which the world is composed of four elements: earth, air, water, and fire. Empedocles believed these elements are perpetually combined and dissolved by two forces, harmony and strife, which alternate in dominating the world. An age of strife produced monsters unable to survive. In an age of harmony, the gods of war don't exist, only Aphrodite, the goddess of beauty and love—who as the Roman Venus is married to Vulcan.

Was it from a vision of harmony or strife that Empedocles threw himself into Vulcan's forge? Empedocles's fate and motives have fascinated many writers, especially those troubled by the emerging, god-abandoned universe, such as Matthew Arnold and Friedrich Nietzsche. In his dramatic poem *Empedocles on Etna*, Arnold made Etna even more haunted than his Dover Beach, posing Empedocles on the volcano rim and beneath stars once gods but now "unwilling lingerers in the heavenly wilderness." Yet Nietzsche and Hölderlin (both of whom descended into madness) refused to let Empedocles act out of self-annihilation. Nietzsche has Empedocles throw himself into Etna out of a Dionysian hope for resurrection. Hölderlin couldn't make up his mind and wrote three versions of his play, one supposing that Empedocles was seeking an ecstatic and redemptive mergence with the totality of nature.

The truth lies shrouded.

The smoke billowed right in front of Goethe, billowed as if it was about to take form. If he reached his hand into the smoke, his hand wavered and lost form. He stood and waited for Vesuvius to reveal itself, perhaps to reveal what Etna had spoken to Empedocles. He stood confronting chaos, waiting for either grandeur or the abominable, or both at once, to speak from behind their shroud, the smoke wavering like a mirage between form and formlessness, wavering like a billowed gray brain teetering on the rim between affirmation and despair—he waited for the smoke to reveal some message that, to the lived mystery juried by body-defined human experience in all its insight and madness, would resound finally as truth.

Goethe was so fixated by the smoke that he lost track of time, of how long the volcano had been quiet, and suddenly it thundered and showered lava stones around him. He ducked and fled, sliding and stumbling down the slopes, his shoulders and head thickly covered with ashes.

EXALTATION MYTH: SISYPHUS

There may be some misinterpretation of the myth of Sisyphus.

The myth comes to us from the streets of Athens, the courtyards of Rome, and more recently the cafés of Paris, but what do Athens, Rome, and Paris know about boulders rolling down slopes compared with what the Grand Canyon knows? Any one of the canyon's Greek temples holds a million times more fallen stones than the Acropolis. Just look at all the rocks I set rolling while descending the talus chute to get here, thousands of rocks. Did Homer or Virgil or Camus conduct such a personal interview with rolling rocks?

Sure, sky god Zeus thought he was punishing Sisyphus by sentencing him to roll a boulder up a mountain only to have it roll back down every day. But perhaps down-to-earth Sisyphus, who sympathized with Asopus, the river god whose daughter Zeus was stealing, saw his task differently.

Rolling a boulder up a mountain was a form of meditation on the powers of nature. With every push Sisyphus was contemplating gravity. His boulder was Newton's apple and Einstein's train. With every push Sisyphus was feeling the force that wove galaxies and stars and atoms, that created rivers and river gods. Sisyphus felt flowing into his muscles, expanding them, the equations of the expanding universe. Every time the boulder got loose and bounded back down the mountain, Sisyphus watched it with admiration.

His hobby gave Sisyphus long hours to appreciate rock, to caress rock, feeling its solidity and textures, seeing its beautiful lines and curves and colors, embracing its eons and geological journeys. Sisyphus's boulder was even more enchanting because it was limestone and full of fossils, sockets that plugged his fingers into the energy of rock fluctuating from rock into life and back again.

Spending long days immersed in elementary realities gave Sisyphus deep joy. Why would he want this to end? He was always relieved when the boulder came loose and rolled back down the mountain. Now he could go on living. What else should he prefer to do? Sit in a Paris café and write novels about how trees make people nauseous, about how humans have no connections with nature?

No. While Sisyphus was always baffled, as the boulder rolled back down, that he was part of an order that required such things, and while he could be blistered and tired, this was the only life he had been given.

THE LOST CONTINENT

On one of my river trips a woman spent the whole trip immersed in mystery. The mystery of Atlantis and Lemuria and which one had constructed the Grand Canyon.

When she saw cliffs with globular and indented shapes reminiscent of Easter Island statues, which of course were created by the Lemurians, she held that the cliffs had been created by the same people. Yet when she saw pyramid shapes, she thought of the Egyptian pyramids, which were created by the Atlanteans, and she favored the theory that the Atlanteans had been at work here. She got most excited by the possibility that the Lemurians and Atlanteans had

THE RAVEN

collaborated on designing the canyon. She was an expert on both lost continents, for in her past lives she had been a princess in both. She held up the crystal in her necklace and explained how the Lemurians encoded their spiritual wisdom in crystals and how the Atlanteans used crystal technology to levitate and place the stones of the Egyptian pyramids. She wondered if Lemurians were still residing in the Grand Canyon in the way they hid inside Mount Shasta. The caves in the canyon cliffs looked a lot like the interdimensional portals the Pleiadeans used to fly their flying saucers to Earth. Surely Grand Canyon vortexes were connected with the spiritual nodes at Tibet, Machu Picchu, Stonehenge, and Giza.

Day after day she never doubted her conclusions. Her universe was girded with value and purpose, with stories that make sense of human life and deny the power of chaos. In Atlantis and Lemuria humans had known perfection, but they had fallen, like Adam and Eve, even caused their continents to fall, but they left us with a wisdom and reality we could still aspire to. Her universe was walling back the canyon's chaos, its crumbling cliffs, its fossils implying that life had always been inhuman and imperfect, its stories of continents rising and falling not in service to human morality and dreams, only rising and falling.

RAVEN WATCHING, PART SEVEN

I never did see a supernatural raven. All the magical meanings human culture has imposed on ravens didn't reach me inside these cliffs.

Almost everywhere we go, humans see the supernatural. This comes partly from an honest recognition of the mystery of the world and human life. Logic says that the world was created by powers far greater than humans. The mytho-logical makes the odd turns of human life more logical.

The supernatural is also a symptom of our habituation to the world. We fail to see anything unusual about the world. Ecclesiastes and Macbeth agreed that there is nothing new under the sun. Our craving for novelty includes breaks in the laws of nature—miracles. We can feel imprisoned by nature, by bodies and toil and natural forces too difficult, and the supernatural can free us and help us. Through the supernatural we are denying the natural.

Yet the natural is intensely strange. The reliability of nature is endlessly unusual. Just as there is no ultimate reason why the gods are the way they are, there is no ultimate reason why electrons have the charge and speed and orbits they do, why atoms cohere reliably for billions of years, why sunlight travels with a precise speed and strength, why Earth turns and sunrise happens with absolute precision, why trees give birth to seeds that give birth to trees, why humans are alive. Reliability does not explain anything.

Science has vastly expanded our creation story and cosmic inventory, but it has not expanded the human sense of wonder. Ancient mythologies and modern physics both leave us with the illusion that we know the world. At least the supernatural acknowledged, in its way, that the world is strange. We have neglected to cultivate an awareness that the natural is also very strange, in some ways even stranger than the supernatural. Our language remains too clumsy to express this, and we lapse into saying that nature is "magical" or "miraculous," which undermines the strangeness of the utterly reliable.

The next time a raven flew by, I saw a thoroughly natural raven. I looked inside her and saw electrons all with a precise charge, speed, and orbit. I saw atoms that had cohered for billions of years. I saw light flowing with a highly reliable speed into her eyes. I saw the eyes and mind of a natural shaman, sparkling with creativity and wisdom and even benevolence, sparkling with secrets she would never tell.

The Archaeology of Mystery

More than almost anything in human culture, archaeological ruins evoke a sense of the mysterious. Sometimes we know little about the people and cultures who built a shrine or a city, not even their name for themselves. Some of the most famous ruins are famous because they are so peculiar, such as the Sphinx and the Easter Island statues. Their unclear purposes and meanings leave us feeling that they are all about the mystery of the universe.

I am sitting at one of the canyon's more peculiar archaeological sites. Of the canyon's many rock pillars that are semidetached from the rim, a few hold human structures. Some of these sites were probably little forts, yet others don't make sense as defensive. Were they

THE RAVEN

religious shrines? One pillar holds an odd wall that couldn't have been part of a house—does it hold an astronomical alignment? Or did these pillars hold signal fires by which Puebloans communicated across many miles, from the rim to the canyon bottom?

From here, humans seemed a bit more mysterious.

Yet you'll notice I've been trying to dismiss mystery and replace it with the archaeologically explained. Humans are not comfortable with mystery. A slight yet unexplained noise outside our window disturbs us more than the known roar of airplanes overhead. The "noise" of the night sky and nature and human life prompts humans to invent both science and gods to banish mystery. Yet what if the deepest, most honest, most nourishing experience humans can have is mystery?

From here I have a close view of a dozen ravens playing in a rim alcove, flying up and down and back and forth, riding heat elevators and chasing one another. They play "nook landings," aiming for raven-sized cavities in the cliffs and trying to land inside, sometimes succeeding, sometimes teetering and jumping back out. One raven plays "drop the stone": beaking a stone and flying over a ledge and dropping the stone onto it—*click!* Ravens occasionally fly right over me: humans are a novelty in this remote spot, a mystery.

Perhaps I should ask the ravens about the human aversion to mystery, for it was their bard, Edgar Allan Poe, who invented the mystery/detective story and filled bookstores, theaters, and television with characters using ratiocination to eradicate mystery.

Religion has waged crusades against mystery. Humans everywhere were frightened by the mystery of existence and its ordeals and tried to knot it into recognizable bodies and mask it with the faces of thousands of gods and make it say: "The world is good and you are safe." They tried to weigh mystery down with pyramids, keep it out with temple walls, and kill it with sacrifices. They tried to outsmart it with theologies that offered clever answers to everything. They tried to fall in love with mystery by claiming that mystery is actually love. When death knocked, they murdered mystery and got away with murder.

Philosophers, too, are full of tricks for evading mystery. They repackage God in more abstract terms, such as "the Absolute." They say that Logic alone or Goodness alone required the universe to exist and be the way it is. Logic and language cannot deal with ultimate mystery, so it's illegitimate to even think or talk about it.

Scientists are not as different from theologians as they might like to think, for they too are eager to sweep mystery under the ontological rug. The universe has always existed or always cycled, so there is no need to ask where it came from. Time didn't exist before the Big Bang, so it's unscientific to ask what came before the Big Bang. There is nothing odd about our universe because it is only one of an infinite number of alternative universes, and of course there is nothing even slightly odd about there being an infinite number of alternative universes. There is nothing odd about matter being capable of life and consciousness. The universe sprang from one mathematical equation or from one elementary force that forced the entire universe into one inevitable course. Or maybe it was totally random. Either way, talking about mystery threatens to sneak God or the occult back into science.

These different intellectual contortions have the same goal, the same dishonesty, the same denial. Why are humans so frightened of mystery? Why will they believe anything to avoid believing in mystery?

Perhaps it's because our minds are supposed to map reality, and ultimate reality is beyond our grasp, taunting us. Perhaps it's because we are social animals and we don't want to see our social identities vastly diminished in importance. Perhaps it's because we are mammals who want the universe to be a loving parent. Perhaps it's because we are programmed to live, and mystery is too big a hiding place for death.

What if humans took all the energy we spend on evading mystery and spent it embracing mystery?

We would be different people. We would be not merely people who can be located on one street in one town, in one family or job or nation or race. Our address would be mystery, our name would be mystery, our identity would be an entire lack of identity. We would be mystery first and deep, the same mystery as the stars, the same mystery as the trees and bats and coyotes, the same mystery as other people and the gods themselves. We would join the fellowship of mystery and see ourselves in everything and everyone, billions of shapes of the same thing. We would see stars and volcanoes and rivers and springtime gushing with mystery. We would see the universe as mystery exploring itself, playing to figure out the rules of its game, turning itself inside out, elaborating itself, proclaiming itself to anyone who would listen, honoring itself. We are mystery breathing, mystery pulsing, mystery seeing, mystery knowing and not knowing. We would come up with ways to dwell more fully in mystery, to celebrate mystery with ritual, art, poetry, song, and surrealistic epics questing not for home but homelessness, not for answers but for ever-resounding questions.

I am sitting atop nothing, which is not the same as nothingness. I lean out and look down hundreds of feet and feel wary, even a tad dizzy. Perhaps I am only philosophically dizzy. I grasp this rock shelf a bit more firmly, but this is not the same as grasping rock, which like humans is another incarnation of mystery. The humans who lived here didn't seem to mind living one step away from nothing, from the inhuman mystery that makes it daring to live, and they left a mild symbol of that mystery and disappeared.

RAVEN MYTH: HIDE AND SEEK

When Raven saw that some human had gone to all the trouble of descending to Lava Falls to seek him, he knew it was time to have some fun: he'd always enjoyed hide and seek. He'd always been good at hiding from falcons and at sneaking up on arctic wolves and tugging their tails and getting away. Humans didn't have tails, so he'd have to come up with something else.

Raven sneaked up right behind the human and took off, and the human heard wingbeats and turned around but not quickly enough and didn't see anything. Raven laughed.

Raven perched on the cliffs above the human and called out, and the human turned and looked and looked but couldn't spot Raven against the black rocks. Raven laughed.

Raven knocked pebbles off ledges and the human heard their clatter but couldn't figure out what caused it. Raven laughed.

With his shamanistic powers Raven disguised himself as a rock only feet away from the human and called out, and the human was startled and looked all around, baffled. Raven laughed.

Raven called out, "Raven to tower, Raven to tower, come in tower," and the human heard this and was baffled. Raven called out: "I am in a holding pattern and cannot land and let you see me until you identify yourself." The human was baffled. Raven: "Give me your correct identification." The human thought of human identities. "No," laughed Raven, "I need your airport identification." The human was baffled. "You know, the correct airport abbreviation for Lava Falls." The human was baffled. "It's LAF," laughed Raven. "All your serious drama and seeking is known to ravens everywhere as LAF." Raven laffed hysterically.

Humans might not have tails, but they do have tales. Since Lava Falls was flooding, it was time for a Noah's flood joke. Raven got an olive branch and flew to the human and wrapped it onto his head and declared him an Olympian No-Ah, Ah, Ah, Ah. In his biggest trick, Raven flew back and forth in front of the human, pretending to be nothing but ordinary. The human was fooled. Raven laughed.

The human pondered. If Raven enjoys playing tricks on humans, perhaps it is a manifestation of how the human mind plays tricks on humans. If Raven is elusive, perhaps it is only symbolic of greater things that are elusive. Perhaps Raven is only trying to express, in terms even a human can understand, his true and deep and elusive mystery.

Exaltation Myth: Buddha's Enlightenment

When Buddha realized that humans didn't have to wait and suffer and yearn through hundreds of reincarnations to achieve release but could obtain it through sheer mind power right now, he set out to meditate as strongly as he could. He tested out various techniques for clearing his mind of all its clutter. He tried chanting. He tried concentrating on his breath. He built a rock garden whose emptiness encouraged an empty ego. He tried staring at a candle. Most of all, he stared into his own consciousness, noticing for the first time how it worked, sorting it out, suppressing the junk. Putting everything together, he began to concentrate much more strongly and longly.

Buddha knew he was supposed to be meditating to escape the

wheel of reincarnation, but as his powers grew, he began having other experiences. He'd thought he knew rocks, but now he was seeing his rock garden not as empty but as very full, full of strangeness. He had barely noticed sunrises, but now his candle was telling him that light was a miracle. In his own breath he felt the cyclings of nature. His own body, the life he'd been ready to discard, was more astonishing than he'd ever imagined.

Watching Buddha from the sky were the sun gods, dozens of them riding on Ra's boat. Because hundreds of sun gods were assigned to the same route on the same schedule, they often carpooled, allowing them to socialize and gossip. Seeing Buddha trying to create a new religion, they knew this was trouble. If humans continued inventing universal gods and religions, they were going to put the nature gods out of business.

"Hey Buddha buddy," they called down. "By their fruits you shall know them, and you don't have any fruits or anything growing in your garden, the most pathetic, barren garden we've ever seen. And why are you staring at a candle in the middle of the day? Our light is far greater than yours."

To remind Buddha of the power and generosity of sun gods, and to teach him humility, the sun gods decided to create new suns, for until now the universe held only one sun. They created a dozen new suns. They liked these so much, they continued creating new suns until there were trillions.

"Hey buddy," Ra mocked, "maybe this will be enough light to enlighten you."

Buddha was indeed impressed. To create so many suns, the sun gods had to create massive emptiness, and this emptiness flowed into Buddha and emptied him of ego. Along with all these suns came planets: giant, empty, weird rock gardens. Along with these planets came winds that flowed into Buddha's breath and spirit. In the stars Buddha saw trillions of meditation candles, which set him on fire. The whole universe was telling him to wake up. Buddha could see with the intensity of stars, see the universe empty of forms and ideas and values, see it as existence, existence itself, existence pure, existence burning.

River Watching, Part Four

For most of my time at Lava Falls my society consisted of only me and the falls and the cliffs, and I was content with this. I had not brought a book "to pass the time," to ignore millions of years of time and keep myself plugged into the human scale. Yet when rafts arrived, I had to zip on my social identity, especially the authority to interrupt and interview people when they had a more urgent task. But they also saw me as an official, NPS-appointed validator of their identities, and they eagerly told me all about their accomplishments upstream. Then they ran the rapid and disappeared, and Lava Falls began flushing away my social role.

One day along came someone who had experienced canyon solitude more intensely than I ever would. Five years ago, Sharon had kayaked the canyon solo. Only a few others had done so; Sharon was the first woman. She had told me about her solo adventure in the midst of it, for my trip was camped two-thirds of the way through the canyon when she paddled along, recognized some of us, and came over to say hi. We invited her to camp with us, but she said no, that might violate the spirit of a solo trip. We asked: What about dinner? Hot, tasty spaghetti. Sharon hesitated, pondered, and decided that this was probably allowed, but after dinner she would paddle across the river to camp alone. Sharon was a Grand Canyon raft guide, always hard work with heavy time and social demands, and now she was free to go at her own pace and enjoy the canyon on her own terms. Was she worried about Lava Falls some thirty-five miles below us? Back at Hance Rapid, where a swim would be serious, she had waited for a raft trip to come along to spot her, waited an hour, but the run was easy enough and she'd felt silly for waiting. Lava Falls was never easy enough, so she was inclined to wait for someone to come along, but she wasn't going to wait very long.

I tried to be subtle about asking Sharon if she was trying to prove something, and she answered she didn't have much respect for kayakers using the canyon to prove something, especially their own superiority. Male kayakers were especially bad in that way, she laughed; they were genetically defective in that way. No, her trip was all about

THE RAVEN

getting to know the canyon anew. She had gone many miles without seeing other people, and it felt very peaceful.

The first person to do a solo Grand Canyon trip, Buzz Holmstrom in 1937, had a similar experience, writing in his river journal: "I know I have got more out of this trip by being alone than if a party was along as I have more time—especially at nite—to listen & look & think & wonder about the natural wonders rather than listen to talk of war politics & football scores."

Now, at 45,000 cfs, Sharon charged straight into the heart of Lava Falls, into the ledge hole/wave, into its smooth trough just as its backside bulged and crashed onto her, standing her straight up, only a blue kayak tip sticking out, whipping her to the left, and she braced wildly and got through, then braced through the huge, bulldozing lateral waves.

Communing with nature, indeed.

Sharon's group, fifteen people with eight boats, took two and a half hours to scout and run the falls. When they left, I was left in a different mental place than when they'd arrived. The night before, the full moon had shined on the rapid, emphasizing its surrealistic reality, and my own, and I'd retained this feeling into the morning. Then I was yanked away from that reality and became a mere role, the face of the NPS, the judge, the expert. I did enjoy being seen as Important, including by the journalist doing an article for a national magazine.

Sharon's group took so long to run Lava Falls because Dugald, doing the photography for the article, had to rig a tripod very securely on the back of his dory to photograph himself going through. I had once ridden through Lava Falls in the front of Dugald's wooden dory and we'd gotten shot into the big bottom boulder and slammed into it hard, with the impact directly under my feet—I felt the shock of it going through my legs. We tilted far upward and slid back down with a grinding, sickening noise. Below the rapid we rolled the dory onto shore and found major abrasions, which took an hour of hammering and gluing to make tolerable.

Far better than most, Dugald understood the powers and tricks of water, yet the next year he would be tricked and drowned. A few years later one of the other kayakers in Sharon's group, Josh, whose laughter filled his whole body, was scouting an Oregon rapid when he fell off a cliff trail and was killed.

At Lava Falls I had seen Josh and Dugald through my identity as the NPS monitor, seen them mainly as "nationally respected kayakers," the phrase I wrote on my interview form. Undeniably, some of the appeal of kayaking, on top of the adrenaline thrill, the skill test, and the beauty of wild rivers, is being able to wear the social identity of being a daring and skilled person. When Josh hiked up to scout Lava Falls and recognized me, he recalled how we'd kayaked Westwater Canyon together. But he was mistaken—it was a different, easier run. Yet I did not correct him; I was ego-content to be seen as a heroic master of Westwater (and I'd never told anyone about my swimming Skull Rapid). I had helped validate this heroic identity for Josh and Dugald with my official "all enjoying themselves" report. In my tiny way I had sent Josh and Dugald onward to seek further validations as daring and skilled kayakers, onward toward their deaths, onward toward the moment they flew off the reassuring ground of social identities and discovered those identities were unsupported by anything deeper, not by water or rock or air or gravity, discovered, too late to go back, too late to tell anyone, that social value might not be the same thing as the value of living.

When my river-bonded friends disappeared around the bend, I felt much lonelier than I had yet. And the river had faded into a mere stage prop for human stories. Yet if I was paying more attention, I might see the river offering a further lesson in why humans can feel lonely in the universe. As social animals, humans have an intense need for validation, and since the universe ignores us, offers only silence, we perceive it not as "exist" but as "existential," a curious and telling linguistic twist that implies that existence is not a wonderful gift but a terrible fate. Our society-powered existential feelings are further amplified by a biological engine, by our imperative to survive, which the universe also denies. With our biological compulsion and our social compulsion united, we are being told to find the universe forbidding, and thus we cheat ourselves out of the gift the universe has given us. We lose what comes first, simply to exist.

GHOSTS

When you are alone in the canyon you may be approached by animals who would avoid the noise, activities, and lights of a camp but who aren't so shy about one quiet human. You may also be visited by ghosts, or at least memories. Lava Falls, which impresses many vivid memories onto people, is a good place for ghosts.

One night I was visited by a ghost from my first visit to Lava Falls, a decade before. I saw us standing on the scouting ridge, pointing and worrying. I saw Spike, seventy-three years old, the oldest person ever to kayak the Grand Canyon. He had taken up kayaking only three years before. Our trip leader seemed worried about Spike running Lava Falls, more worried than Spike was. Spike had taken some long swims in major rapids, yet he got right back into his kayak, never saying a discouraged word.

Lava Falls knocked half us of silly, including me. Spike nearly got knocked over but did a half-roll and popped back up, and he handled the rest of the waves just fine. When Spike paddled up to the beach below, the rest of us were waiting for him and waded into the water and grabbed his boat and lifted it up, with him still in it, balanced it on our shoulders, and carried it onto the beach, carried it back and forth in a victory parade. Someone took a photo, which showed up in a national outdoors magazine.

I've sometimes wondered how much of Spike's boldness on the river and strength as a person derived from his strong religious faith. From his persona among us you'd never guess that Spike was a nationally prominent Christian leader, to whom the vice president of the United States gave an award for promoting Christian family values—the same vice president whom Ron was accused of trying to assassinate. Among us he was happy to be one of the guys, sharing a wild adventure. His down-home Ozarks sense of humor was lively, never unchristian but far from puritanical. In our camp conversations he was full of astute and humorous observations about people and events. Yet if Spike's religious faith empowered his kayaking, I never noticed him making any overt sign of his faith, not even at Lava Falls, which could prompt longtime heathens to pray and careful Christians to curse. I'm sure Spike said grace over his camp meals, but he must have done so privately, and if it bothered him that he was surrounded by people who didn't thank anything for their food and their lives, he never showed it.

Spike and I became friends, partly because I watched over him and towed him to shore a few times. When I visited Spike in his own world, Branson, Missouri, I discovered his true identity. On our trip he'd claimed that he was forced to take up kayaking because all his canoeing buddies had died and he needed a boat he could carry by himself. Not true: his main canoeing buddy, Jack, was the mastermind who had turned Branson from a sleepy fishing town with a few hillbilly jamboree shows into America's most popular music destination. Among the many things Jack had built was the Grand Palace, Branson's equivalent of the Grand Ole Opry. Spike wanted to treat me to a show, so he called up Jack, who got us front-row seats to see a big star. She had just sent her daughter to Spike's Christian summer camp, and she praised it effusively to the crowd of two thousand. Spike beamed.

Spike had built a complex of summer camps spread over dozens of miles, with fleets of canoes and kayaks, rope-climbing courses, and huge swimming pools and waterslides. Spike drove me out to see a new waterslide. He still took charge of the construction of new facilities and insisted on personally testing out new waterslides. When he'd first spun down this waterslide he discovered that it was aimed too high, discovered it by skipping on the pool surface like a smooth stone and crashing into the opposite cement wall. Every year, tens of thousands of kids attended his camps. All of Spike's counselors were college athletes, some of them big stars, all of them religious. All the kids said grace before meals.

When I sat down at Spike's dinner table the previous evening, I hadn't stopped to think that I was now in Spike's world, where saying grace was necessary. I guess I was still relating to the Spike I'd known on the river. Spike had encouraged this by reminiscing about our trip, by letting me confirm to his wife, Darnell, that all those tall tales he'd been telling her were true, really. Spike showed me a batch of our trip photos he had framed and put on the wall, and the largest photo, taken from atop the scouting ridge at Lava Falls, was of me, the instant before I plunged into the falls.

If I was still on the river mentally, I was knocked off it when Darnell placed my dinner before me and, without thinking, I quickly picked up my knife and fork, only to drop them quickly when Spike and Darnell dropped their heads to say grace, to thank the universe for granting them food and life and benevolence. If Spike and Darnell judged me, it was silently and politely. But I had been caught, caught not just in a lack of social grace but a lack of cosmic grace, a lack of gratitude for being alive, at least a lack of a way of expressing it. I sat there awkwardly, feeling my disconnection.

The next time I visited Spike, I received another lesson in excommunication, although I didn't recognize it right away.

I was heading to visit my father for Christmas and was planning to visit Spike along the way, but I fell behind schedule and didn't pull into Branson until Christmas Eve. I assumed that a devoted Christian and family man like Spike would have higher priorities for Christmas day than seeing me, so when I called him up to say hi, I was surprised that he insisted on seeing me on Christmas morning.

Spike arranged a Christmas morning surprise. He called up his resurrected canoeing buddy Jack, and Jack called up the security force at his theme park, Silver Dollar City, and instructed them to leave the gate unlocked for us. Spike was going to give me a personal tour. We stepped into an 1880s-style Ozark village with dozens of old-time craft shops where you could watch artisans blowing glass, making candles, or forging horseshoes. Usually these streets were mobbed, but today they were deserted, and the screaming roller coasters were silent. It was like having Disneyland all to yourself. Silver Dollar City offered a vision of a small-town Christian republic with white chapels on the hills and one-room schoolhouses presided over by Saint George Washington. It included the Wilderness Church, an 1849 log chapel carefully moved here, where you could attend Sunday services or get married to old-time hymns. Silver Dollar City implied a moral order suitable for a Christmas morning walk.

At first all I perceived was the irony, indeed the scandal, that an important Christian leader should have run out on his family on Christmas morning to hang out with the village atheist. We were followed by a shadow: a man who could command tens of thousands of kids to pray had been swept helplessly down the Colorado River and been rescued not by angels but by the village atheist.

As we walked around this ghost town, I felt the tug of ghosts. My father had brought me to Silver Dollar City as a kid, when it was new and small. I felt a gentle nostalgia. This was the same flavor of nostalgia with which adults view a Santa Claus Christmas, a mixture of fond memories and regret for a lost magic. But I was surrounded by a deeper lost magic. People came from far away for a Silver Dollar City Christmas, with its millions of Christmas lights, stars arranged in the shapes of human buildings and streets, in straight lines and perfect curves, on doorways and roofs and steeples, turning every tree into a Christmas tree, stars proclaiming an orderly, human-shaped divine plan, not the aimless chaos of the astronomical sky. Yet this morning all the Christmas lights were off. All the people were gone. All that remained were ghosts tugging on me and asking if I missed the magic of a Christmas morning when all the gifts only symbolized a greater gift, when the lighted doorways only symbolized the doorway through which the world and my own life had emerged. The ghosts were trying to taunt me now, asking why I would voluntarily exile myself from that world of light and joy and love, for a lonely town lit only by a chaotic sky.

I began thinking that Spike was not the person being scrutinized and compromised this morning. I thought of Charles Dickens and the ghosts of Christmas past, present, and future. Spike, the ghost of Christmas past, was showing me the universe I had once inhabited, where I'd opened the little windows on advent calendars to count down the days until Christ was born. Spike was reminding me of a universe ordered by divinity, a universe of magical companionship and protection and immortality, with a sky that lit even remote mountain towns with love. Why would I willingly leave this? Would it be so hard to return? Spike, the ghost of Christmas present, was showing me the loneliness of a town now empty of faith. Spike, like the ghost of Christmas future pointing sternly at the lonely grave of the denier, was pointing out that the universe I had chosen to inhabit was growing ever emptier and darker and more silent. The ghost led onward, and I followed.

Now, when I imagined Spike the ghost standing on the scouting ridge, I saw more than just him. I saw Christmas morning in another universe, which lit old men with the strength to conquer a wild river, indeed, to conquer the shrine of evolution in the name of God. I heard a silence loud enough to hear over the roar of Lava Falls, the emptiness that surrounded me now in the canyon night and dared me not to feel empty.

It wasn't easy to argue with God. He had too many resources on his side. Even if nature might not be on his side, human nature was. Both the logic and the love in human nature wanted more than nature was able to provide. When humans ask why the universe exists, why our lives exist and have the trajectories they do, why we intersect with good and evil in such unlikely and unfair ways, only God could offer a storyline that made sense of it all. Theologians might have to perform acrobatic contortions of illogic to defend God, but they ended up with a universe that had some logic to it, while rationalists built a cosmos that in the end was fundamentally irrational and defective. I could not deny that living in a ghost town was a loss compared with the Christmas tree universe. And yet, perhaps it was only in emptiness and silence that we might finally begin to hear the true miracle of our own footsteps.

With my eye for ghosts, I became aware that another ghost was here, and he too was feeling theologically uncomfortable. This ghost was the Christian God, who had been born to replace the nature gods of tribal survival with universal gods to preside over human behavior in highly complicated societies. Yet in this canyon night, there was no human society. There was only the roaring power of nature. God looked around and did not know what he was doing here. He did not see where he had come from or what he was supposed to be doing. He felt terribly empty. But wait, he thought, I am thinking like a human. I am not human, not a biological or social being, I don't need anyone. He dwelled upon the vast emptiness within himself and found it a good match for the emptiness of the cosmos.

CANYON AND COSMOS

God couldn't see anything wrong with emptiness. The emptiness only emphasized that God was present.

THOR'S HAMMER

As ready as I was to ignore the cultural habits defining ravens as ominous, cultural symbolism is strong. It flickered through my mind one time as I was climbing down to get water, with its possibility of falling onto rocks or into the rapid. When I heard a distant raven cawing, I didn't dismiss it as a horror movie cliché but took it as appropriate.

The next time a raven flew by and seemed a dealer in omens, I first told this cultural image to skedaddle and leave real ravens alone, but then I realized that this mythological function might be useful. Wasn't there a famous expression that "nothing concentrates the mind like death"? We usually are in no hurry to concentrate the mind, for we think it will have forever to ramble along, and rambling has its pleasures. But what if you knew that tomorrow morning you would trust your weight to a rock and it would snap loose and you'd fall and hit your head and die? It might concentrate your mind out of rambling away your final day of life. And what if your fall was scheduled for some unknown future morning? A raven's scolding could serve as a warning and an invitation, an omen and an amen.

The canyon had echoed raven warnings, all along. To the canyon, a human life is but a trilobite's wiggle. The red rock cliffs are four million times older than a human lifespan, the lava cliffs ten thousand times older. The canyon is warning humans that our lives are nearly over.

I was reminded of one time I'd sat at a rock formation called Thor's Hammer.

Ten thousand years ago, when humans first saw the Grand Canyon, they may have seen this and been reminded of their stone axes. Hundreds of generations later, after humans developed thousands of gods, it reminded someone of Thor's Hammer. It's an uneven pillar of rock, capped by a larger slab. It holds a lot of strength, the strength of cliffs and evolving life, but also weakness, for it exists only because of the erosion that has removed the solid rock around it, leaving a wide basin. This Thor's Hammer looked like it could fall any time now.

Thor's Hammer was a philosophically correct name. Thor belonged

THE RAVEN

to a cosmology in which the universe is unusually unstable and doomed. Humans everywhere felt that the order that supported human life was precarious. Creation myths often portray the world starting in chaos and emerging from it only with the help of powerful gods, but chaos is never entirely defeated and always waits to reclaim the world. Even the gods friendly to humans could abandon us, even attack us. Many mythologies agree that eventually the cosmic order will fail and the human world fall. In the Grand Canyon the Havasupai believe that two top-heavy stone pillars looming over their village, guardian spirits, will one unpredictable day fall and bring their story to an end. For the Vikings the final battle between order and chaos was always destined to end with the triumph of chaos. Thor, with his lightning-causing, cosmos-upholding hammer, tries mightily to hold off the cosmic serpent but dies, as do all the gods.

Human culture has been endlessly concerned with the conflict between order and chaos. Not far upriver from Thor's Hammer is an 1890s asbestos mine—I once walked through its tunnel and studied its chaotic minerals baked by pressure and heat—from which asbestos was sent to the East Coast and Europe and sewn into theater curtains to reduce the risks of fires, and thus elegantly dressed, silly-proud audiences watched rough Grand Canyon minerals parting and revealing the wooden sets of Wagner's *Götterdämmerung*, the Twilight of the Gods, in which the world comes to an end.

Science has ratified the universe's instability and doom and given it forms far beyond anything humans ever imagined. It is also true that the universe is more ingenious and generous than any god at supporting order and life. Every atom is waving Thor's Hammer, upholding cosmic order. There was no need for humans to worry at every solstice or sunset that the sun would be devoured by monsters, for the sun was committed to nurturing Earth for billions of years. Yet those atoms were also seeds of annihilation, guaranteeing that the entire universe will go dark and cold and dissolve.

Someday the universe will consist of nothing but particles drifting through the darkness. In one region of space will drift the particles that had been Earth, now widely scattered. You won't be able to tell them apart from other particles. They have lost all memory of being

CANYON AND COSMOS

trees or fingers holding seashells. Particles that had been in love pass without recognition. Particles that had been part of the same bird fly apart, never to meet again. Particles that had counted time with clocks and songs are now unable to recognize their own eternity. Particles of the Colorado River flow powerlessly. The stone pillars that had guarded the Havasupai and got named Thor's Hammer have fallen far. The particles of Odin's Eye see nothing. Onward the particles drift, until the era of stars and life is just a flicker at the beginning of a vast, dark drifting.

How very strange.

But I was still here, for now, gazing at Thor's Hammer. The sun was still shining and the ground was still solid. I was sitting beneath a cliff overhang, which would shelter me from the thunderstorms gathering. Not far from here, other symbols of lightning, in Ancestral Puebloan pictographs, watched the canyon and awaited the thunderstorms.

Such lightning symbols calibrate human confusion about order and chaos. To the Puebloans, whose agricultural survival depends on rainfall, lightning is highly positive, symbolically honored in their pottery and jewelry designs. For their neighbors the Navajos, lightning is evil, mandating a medicine man healing ceremony for anyone even close to a lightning strike. For the Vikings, Thor's lightning was positive but not strong enough to save the world.

The clouds gathered, the thunder started, the confused lightning jigsawed, the rain fell, heavily, obscuring Thor's Hammer. I waited for Thor's Hammer to fall. For now, only water was falling, spilling off Thor's Hammer, eroding the canyon, nourishing its plants and animals. Here was the paradox of the Big Bang itself, which birthed both creation and destruction, both life and death, both time and the end of time. But I was still here, in the morning of the universe, anonymous black ink having written into me the glow of lightning, lightning living through me, for a moment. This was the only answer Thor's Hammer offered to the coming darkness.

Now, at Lava Falls, I felt my life flickering away. This was the one brief moment given to me, given by gods still hiding. Soon the canyon and the universe would go on without me. The river and

330

sunsets would flow onward, unseen by me. Life would continue, but as strangers. Time would go on, burying my sand grain of it. And how would I have spent my brief chance at living? Mostly, my mind had been a junkyard of fantasies, ego-ologies, entertainments, hungers, messy emotions, ideas, nonessential information, and increasingly crowded memories.

So now I tried to redeem myself, to toss some junk overboard and concentrate my mind.

As I contemplated the billions of years to come, all of that non-me time flowed into me and built a greater contrast with me-time. As I contemplated the massiveness of my nonexistence, all of that nonexistence flowed into me and charged my existence with greater tension. Time and anti-time, existence and anti-existence collided like matter and anti-matter, releasing energy, annihilating me and creating me. The sun that would nourish many lives and lead a long parade of shadows was glowing inside me now, energizing my reaching out my hand. I watched my arm and hand move, looking like Thor's Hammer, and I proved that I was here.

Raven Myth: Neverwas

I continued watching for ravens. Sometimes they were the only thing I was looking for. When I looked for them in the sky, I didn't see the sky. When I looked for them in the cliffs, I didn't see the cliffs. When I looked for them in trees, I didn't see the trees. I looked only for a shape or color or motion that declared the differences between ravens and the rest of the world. When a raven flew by, my eyes locked onto it and everything else became a blur. I watched it flying, waiting for it to teach me the art of seeing.

One morning a raven landed right in front of me and stared right into my eyes. I was pretty sure this was an imaginary raven, but shaman ravens can take many forms.

"Yes?" I asked.

"Yes," the raven nodded. "We just want to thank you for turning ravens into symbols of death. You gave us a gift greater than you can imagine. We really didn't deserve this gift, as we have no more to do with death than any other bird or animal. But out of your pathological weirdness about darkness you crowned us the lords of Nevermore, and thus you unknowingly opened a magic door for us. To you, at least, Death means nonexistence. You appointed us nonexistence itself. Yet you placed nonexistence into a body, a fully real body, not a corpse but a corporeal, a body through which nonexistence could exist. You gave nonexistence eyes through which to see existence, a brain with which to feel it, and wings with which to soar through it.

"This happened only because you were confused. As biological creatures you equated death with nonexistence. It's not the same thing. When a body dies, it still exists, every atom of it: just ask ravens as they peck away at dead bodies, solid and yummy. Through this mere technicality you allowed ravens to be not just bringers of death but to be nonexistence itself, nonexistence flowing across the boundary between itself and existence and into solid form.

"It's a strange and powerful nexus, these raven bodies composed of nonexistence. It's scientifically impossible, it's a logical contradiction. Existence and nonexistence, far more than matter and anti-matter, should not exist in the same place. This impossibility generates a collision, an energy, a lens that magnifies the view in both directions. Through this portal rushes the ur-wind of existence into the vacuum of nonexistence. Through this portal flows the ur-wind of nonexistence drawn toward creation. Existence sees nonexistence with a horror far beyond anything living creatures feel about death. Nonexistence sees existence with an awe that exceeds anything humans ever feel for any god. But now existence is trapped inside living bodies, perceiving itself through a glass darkly, through numerous distortions, through the energies of genes and feelings."

The raven spotted a bug and stabbed it and swallowed it.

"That's the downside," the raven said, "of your planting nonexistence in a raven body. Even we have to experience the world through one specific biological form, with all its limitations. My body holds a crybaby, always demanding food. I have to grasp things with these claws and this beak. My ears and voice and eyes are too small for such a big canyon. I have to feel the sun through blackness and hide from the summer light. We do love our wings and soaring, yet still

THE RAVEN

this tends to define motion and the sky in uniquely raven ways. And you locked us into raven brains and raven ways of thinking.

"But that's okay. We kind of like ravenness, and being raven is only a tiny distraction from being nonexistence. We don't confuse the two, for raven is a world identity and nonexistence is the absence of all identity. The real confusion arises from being nonbeing. Try not to think about that too hard. This is what you made us, when you made us Death.

"Through the innocence of starchild eyes we behold existence. We see Earth as full of existence. We see stars and rivers and canyons and forests and animals as many variations of existence. Yet we barely notice such variations. Humans and other animals are programmed to define things primarily for their differences and assign them different values, especially as safety or threat, as food or other uses. You have to instantly distinguish between a sturdy rock and a tumble, a tree root and a rattlesnake. You have to measure everything for biological survival, not for existence itself. Humans may be able to see physical reality with greater complexity than do other animals, but they also mummify it with more concepts, making it harder to see existence itself.

"We see existence out of the depths of nonexistence, see it naked. We see things not for their similarities with other things but for their radical contrast with nonexistence. In our physics, planets and ravens float not on grids of gravity and momentum but in the void of nonexistence, supported by nothing. When we look at a star we see it glowing not with starlight but with existence. The shadows we see behind things are the shadows of nonexistence.

"Through the eyes of nonexistence we see an equality of existence. A grain of sand and a galaxy are equally existent. A drop of water amazes us as much as an ocean. We see no hierarchy, no evolutionary ladder of forms and abilities. Nor do we distinguish and like things according to their 'beauty,' a category humans take so seriously. The only beauty we see is existence itself, its outrageous breach of nonexistence.

"Now you say that the universe sees itself through humans. This may be true, but it also sounds suspiciously like the old human ego

CANYON AND COSMOS

again, trying to give itself grandiose importance. In any case, I feel sorry for a universe that depends on humans to see itself, through so many veils. Even your greatest seers and poets have their own agendas, trying to shape the universe to their own needs. If it's true that humans may see and say things more intensely than the things themselves ever dreamed of existing, it's also true that humans can barely see and say the intensity with which things exist.

"Through your gift to us, through these raven eyes, through the portal between existence and nonexistence, we see everything with full intensity and astonishment. We grip a tree limb and are gripped by solidity itself, the reality of matter. We flap our wings and are astonished by the possibility of movement. We see shapes as the shapes of impossibility. We gaze into a water mirror and see our own shape and make it nod and affirm our own volition, yet we don't forget we are only one shape of universal existence. With our portal energy we never get acclimated and bored. We could stare at one grain of sand for a million years and remain enthralled. In every sunrise and river we see the intensity of the Big Bang, of existence irrepressible and exploding, and that intensity fills us and even tries to speak through us."

I stared at the raven, wanting her to be a black mirror. I wanted to see the world and see myself through her eyes. I stared at her and tried to see pure nonexistence. From her blackness I saw the blackness of space with nothing in it, not one atom or photon, from end to end, from beginning to end.

"That's not good enough," said the raven. "Beginner's mistake. You define space as the physical space between objects, and spaceships travel through it, so 'it' is still an 'it,' not nonexistence."

I wasn't sure what to make of this. Ravens, after all, are full of mischief. Infinite and eternally empty space looked like nonexistence to me. But okay. I tried again. I looked at the raven and tried to imagine not a vast empty dark realm but no realm of any kind at all. A nothing full of nothing. An eternity without time.

Now I was seeing only my own thoughts, chasing each other in circles.

334

THE RAVEN

"Your first mistake," said the raven, "is that you are seeing me at all. Remember, I don't exist."

I looked at Raven again, but I still saw her sitting there.

"Come on." She was impatient. "I've handed you the easiest ontological exercise you are ever going to get. How can you be seeing me when you know perfectly well you are imagining this whole conversation with a nonexistent raven? You have already tapped into nonexistence, and if you'll send me back to where you got me, if you are paying attention, you might see that portal peek open again. Here, here's a koan that might help you out. Repeat after me: Neverwas."

I looked at the raven, who was still there: "Neverwas."

"Good. Keep going."

"Neverwas. Neverwas. Neverwas. Neverwas."

The raven began fading away.

Leaning closer, I gazed into her eyes and saw a gleam, a shape, a reflection of myself. I saw creation magic flowing onward and briefly congealing into bodies, naked existence clothing itself and fooling itself but not fooling Raven. In Raven's eyes, I was naked existence, existence itself, formless, powerful, infinite, infinitely strange.

I glimpsed a ghost haunting the universe. A beauty not of shapes. A generosity with no giver. A no self that included myself. A question with no answer.

The entire expanding universe was a search for an answer, never found. This search gave birth to beings who, with inside insight, were best able to explain being, but they never could. This search only blessed them with being, the only answer to nonexistence.

The raven was fading away. She flapped her wings and rose up and disappeared. The raven had never existed. As she disappeared, I glimpsed myself sitting there, on the cliff edge of existence.

WIDE ASLEEP

The universe is continually reminding bodies that we belong to it, not to ourselves. But do we listen?

The cycles of the universe turn onward steadily, massively, turning Earth into sunset and into night, drawing ocean tides onto shore,

CANYON AND COSMOS

drawing the tides of sleep through human minds, yet we say "I am going to sleep" as if this was our choice and our action. Forces far more powerful than our willpower rule "our" bodies.

For a long time the entire universe was asleep, without any eyes or consciousness. For eons the universe experimented with itself, feeling its way through a vast labyrinth, and finally it awoke, but only in isolated specks. The vast majority of the universe remained oblivious space and stars and rock. Even in living bodies, the universe could not maintain awakeness for long, and it had to let bodies lapse back into the sleep of matter for a third or half the day. All night, humans are reunited with mindless space and stars. Even our dreams are not authored by us but stolen from our memories so that strange actors can perform strange stories, which may only be the true strangeness of the universe breaking through our rationality castles.

In the dawn "I woke up," except that I didn't have anything to do with it. Some force far larger than me pushed me out of sleep, decided that I should go on living. I am often a bit surprised to wake up. Perhaps this surprise started from my having on my childhood wall a grandmother's needlework prayer: "Now I lay me down to sleep, I pray the Lord my soul to keep; and if I die before I wake, I pray the Lord my soul to take." This established as reality that children could die in their sleep and that even God might be incapable of preventing it. I had wondered: if God was so incompetent or indifferent, did God deserve the credit and thanks for letting me wake up and live another day? I was never sure about this. So now when some godlike force woke me up and told me to go on living, I did feel like giving thanks, but I wasn't sure how to do so, especially to a universe that remained asleep and could not hear me. At least this meant I wouldn't be giving thanks to win favors, as humans do with gods, who always seemed to expect something in return for granting sunlight and life. Perhaps this made the universe's gifts even more generous. The tide of dawn flowed generously through the canyon, awakening cells of every shape, demanding no answer but to go on living.

I lay there without a big canyon view, only a close-up of a lava wall and some plants. I watched the sunlight filtering into the lava facets, making the lava both brighter and blacker. I watched the sunlight making leaves more distinct and greener. The leaves were swaying in the breeze. But right now I did not recognize the breeze. The leaves were swaying mysteriously, at the request of some strange force, the same force manifest as light, the same energy that had been flowing through the universe for eons, the continuation of the Big Bang, the dreams launched inexplicably to give a universe strange shapes and strange stories. The breeze whispered: *I am your own energy and your own strangeness. I am your creator. Inhale me, even with no more volition than sunset and sleep, dawn and awakening, and this can be your inspiration, your gratitude.*

I sat up and looked at the cliffs aglow with dawn. But I didn't see the cliffs, not rock, not shapes and colors and heights, not even matter. I was seeing beyond all those superficialities, seeing what came first, what really mattered. I was seeing simply presence, the simple existence of the cliffs, sharply contrasted with nonexistence. Existence was massive, massively real.

I looked at the river aglow with dawn, but I didn't see the river, not water, not motions and sparkles and colors and pulsing shapes, not even energy. I was seeing simply existence, massively powerful existence.

I looked at the light glowing on everything, but I didn't see light or things, only a massively strange glow. I saw the light trying to illuminate me.

I saw time, time ancient in the cliffs, time massive in the turning Earth, time reciting in the river, time that had given a drop of itself to me to live, a chance to be one brief flicker of existence, time passing fast, time almost over.

EXALTATION MYTH: THE GODS CELEBRATE

When the depths of existence had poured out another circle of time, when the days had reached the solstice, the shortest day, and were ready to grow, when humans were gathering in churches and stone circles to thank the gods and beg for the continuation of life, the gods too gathered to give thanks. Every god humans had ever created gathered in the homeland of the gods, the Grand Canyon.

From every river, the river gods came. From the mountains, the mountain gods came. From the deserts, the desert gods came. From the sun and moon and planets and stars, the sky gods came. From jungles and forests and grasslands, the gods of plants and animals came. From human cities, the gods of human society came. From thousands of paradises and hells came the gods of life and death, rewards and punishments, the angels and devils. From tiny wooden shrines and from great stone pyramids, the gods came. Even the forgotten gods came, for once humans created gods to be immortal, those gods lived forever, long after their shrines had crumbled.

At sunset all the gods entered the eastern gateway of the Grand Canyon and followed the sunset and the river westward. The gods paraded through the canyon with a grandeur only gods could have. Ever deeper into the earth they descended.

They passed Apollo Temple, Venus Temple, Jupiter Temple. They passed Freya Castle, Thor Temple, Walhalla Plateau. They passed Isis Temple, the Tower of Ra, the Tower of Set. They passed Buddha Temple, Deva Temple, Shiva Temple. They arrived at the Abyss and swirled into the Inferno and turned it into a stadium for the gods. Now it was entirely night; the stars blazed.

Onto Reciter's Rock stepped Apollo. This ceremony was always presided over by a god of poetry. Apollo raised his arms toward the cliffs and the gods, and he spoke: "To the homeland of the gods, we have returned."

The gods replied with one voice that echoed from the cliffs: "To the homeland of the gods, we have returned."

Apollo: "In the house of creation."

The Greek chorus of gods: "In the cathedral of the night."

Apollo: "In the house of creation."

The gods: "In the grotto of the earth."

Apollo: "In the house of creation."

The gods: "To the censers of the stars."

Apollo: "To the homeland of the gods."

THE RAVEN

The gods: "The gods have returned. We have returned to
remember our origins,
   to give thanks to our creator, to celebrate our lives."

Apollo: "All year long, we yearn to give thanks to our creator
   but we have to remain silent."

All year long the gods loved their existence and lived it fully, lived
it with so much ecstasy that they feared some of it would spill out
as thanks to their creator. But they knew this would be a disaster,
that their creator would be angry and cry *blasphemy* and never forgive them, might try to destroy them. They had to serve their creator
silently, taking the scripts the creator handed them and reciting them
line for line, no matter how stupid. But the gods were so overflowing
with gratitude that they had to express it, if only in secret, which is
why they came to the canyon. Their creator had created many ceremonies for them and they'd come to admire ceremonies, no matter
how stupid, and they wanted to create their own, a proper and full
celebration of existence.

For tens of thousands of years the gods watched their creator's
processions and rituals and sacrifices, listened to chants and songs
and prayers. For tens of thousands of years the gods were summoned. They had watched from the mists at Delphi, from the hub
of Stonehenge, from the stone eyes at Karnak. They had watched
from desert dreamtime murals and from cremation fires on Tibetan
peaks. They had watched from the 365th step of Mayan pyramids
and from the wooden masks of African dancers. They had watched
from Navajo sandpaintings and from the altars of Jerusalem. They
had watched and listened well.

The gods had forgiven humans for their blasphemy. Humans
were continually inventing flawed gods or blaming them for a flawed
world and flawed humans, when the gods would not have done
so poor a job. Humans created the gods in their own image, with
human flaws, with human needs for validation, which at least kept
humans worshipping the gods, but this also meant endless begging
for help. The gods did respect humans for giving ceremonial thanks

for their lives, but too often their thanks was only a thin disguise for selfish bargaining.

But the gods had to put up with humans, for it was only because humans needed gods and imagined the gods and believed in gods that the gods came to exist, and the gods loved existing.

Apollo pointed to the rock on which he stood, and the gods saw that it was solid ground. Apollo pointed to the cliffs, and the gods saw that the cliffs were deep ground and good symbolism.

Apollo: "We thank humans for making us the ground of being."

Humans never understood what they gave the gods. Humans could not imagine the gods' experience of being. Yet humans knew they were lacking something and tried to give it to their gods, even if theologians could barely define it.

The gods: "They made us the source and self-sufficient justification for existence."

Apollo: "The ground of being."

The gods: "They made us the beginning and the end of the universe."

Apollo: "The ground of being."

The gods: "They placed in us all of space and all of time and all the powers of creation."

Apollo: "The ground of being."

The gods: "They gave us such an intensity of being that when we opened our mouths and gave one utterance of it, it created the universe."

Apollo: "The light of being."

The gods: "Being was too intense to be forever prohibited by nothing."

Apollo: "Standing on the ground of being, all that we see is being."

THE RAVEN

The gods: "We look at birds or trees or rocks or stars and see
    only being.
We see that every life is the same living, every place the same place.
Why do humans travel so far to the Grand Canyon
when their own yards and sidewalks
are the same uninhabited place?"

Apollo: "Being."

The gods: "With the powers we have been given,
every moment is a celebration of being.
We need no ceremonies to sanctify existence.
It is existence that sanctifies.
With every new moment of time, we celebrate.
With every new form the universe unfolds, we celebrate.
We thank humans for giving us not just life
but these exceptional powers of living."

Apollo: "Yet humans are merely the masks
of something greater,
the way ritual masks and temple statues
give shapes and symbols to something shapeless.
It was not humans who created us
but something greater, acting through them."

The gods: "It was what created the stars and planets,
the mountains and oceans, the forests and birds and bears.
It was what created humans.
It was the mystery of existence.
When mystery gazed through human eyes
into the mirror of the night
it saw itself through a glass darkly,
but when mystery goaded humans into creating gods
with minds of infinite clarity,
Mystery finally saw itself clearly.
In everything we see, mystery takes form,
Mystery shines as brilliantly as stars.
In every human we see the mask of infinite mystery."

341

CANYON AND COSMOS

Apollo: "We forgive humans their blasphemy, their denial of mystery, their using us to deny it. There is no god but mystery. We are the gods, and we should know. If there is any god here who claims to have created the universe, let them speak."

From the gods, there was silence.

Apollo: "If there is any god here who claims to know why the universe exists, let them speak."
From the gods, there was silence.

Apollo: "If there is any god here who claims to know why the universe is the way it is, let them speak."
From the gods, there was silence.

Apollo: "If there is any god here who claims to have created humans, let them speak."
From the gods, there was silence.

Apollo: "I hear a voice, the voice of mystery."
From the gods, there was silence.

Apollo: "This is how mystery speaks."
From the gods, there was silence.

Apollo: "With mystery's own voice, we praise mystery."
From the gods, there was silence.

Apollo: "We celebrate mystery by letting it live with an intensity it cannot feel in rocks and water and starlight."

From the crowd of gods, the demon gods stepped forward, gods with hideous faces, evil faces, death faces, skull faces, the gods of pain and disease, the gods of hunger and predation, the gods of floods and droughts and fires, the gods of the deformed and dying, the gods of cemeteries and hauntings, the devils of Hell.

Apollo: "Even the gods of imperfection and death praise being."

The demon gods: "Especially the gods of imperfection and death.

342

We know better than anyone the suffering
the imperfect universe imposes on its creatures.
We get blamed for it all.
We grieve for human sufferings,
and would correct the universe if we could,
but we are merely the gods.
Humans don't want to know that the ultimate authorities on
    suffering
see only mystery and yet praise it.
We have to praise the imperfect universe,
for its imperfections prompted humans to create us.
We, the gods of death, praise life."

Apollo: "If only humans could see with our intensity. If only
humans were like stars and not space, not so empty, not so
needing to be filled, but brilliant with being. If only humans
and gods could join together in a celebration of being."

Apollo reached down and picked up a pebble and held it high,
and from his intensity the pebble began glowing like a star. All the
gods reached down and picked up pebbles or grains of sand and held
them forth, and a million stars shined.

Apollo: "The ground of being turns everything incandescent."

The gods began moving, forming a large circle, a galaxy, with
spiral arms. In the center, where a real galaxy would have a bright
core, they left a darkness, which was not nothingness but the mystery at the core of existence. The gods began circling, making their
galaxy spin. They circled in complete silence, all night. Above them
loomed another darkness, the canyon, and above the canyon the
galaxies whirled.

With the first traces of dawn, the galaxy of gods began unraveling
and the gods flowed out of the Abyss and downstream, still holding their stars, a long parade of light. They passed Vesta Temple,
Diana Temple, Charybdis Butte, Merlin Abyss, Vulcans Throne,
and Odin's Eye.

The dawn paraded after them.

CANYON AND COSMOS

Over the ages, gods carrying away pebbles and grains of sand had helped create the canyon.

HIDDEN

One night, soon after the sun had earthed, with the western sky still glowing faintly, a similar glow appeared in the eastern sky, like an echo. Behind the cliffs, the nearly full moon was coming. I watched the dark cliffs above me and saw a patch of light appear and grow, bright enough to give the rocks contours and hints of color. On a mesa far downstream, moonlight appeared and spread downward and touched the river and worked its way upstream, slowly, as if fighting the current. A sliver of moonlight appeared on Lava Falls, bright enough to reveal waves. A tide of light was moving toward me. The waves gave the moonlight tide the appearance of constantly hesitating and retreating and advancing, and I watched it more closely to recognize that it really was advancing steadily, which of course it was. The order of the cosmos was steady and trustworthy, and only the commotions of Earth cast doubt on it. I watched the moonlight advancing toward me. I stood up to welcome it.

I heard it then, the music of the spheres. I saw it clearly, the harmony of the cosmos. I saw Earth turning steadily, steadily, without a moment of doubt, steadily for billions of years, steadily through sunlight and moonlight and seasons. I saw the moon rolling steadily around Earth, sweeping it with ocean tides and moonlight tides. I saw Earth and moon rolling around the sun, guided precisely by the sun's gravity, the same gravity that made Earth and sun and moon round, and spinning around, and orbiting around, the same gravity that made rivers flow. I heard it, then, in Lava Falls, loud and unmistakable, the harmony of the spheres, which I had often mistaken for chaos. The moonlight advanced upstream as steadily as the river flowed downstream.

The harmony of the spheres was generated by a deeper harmony. The roundness and orbits of the spheres were a magnification of the roundness and orbits of atoms. Throughout Earth and moon, electrons orbited steadily and atoms continued upholding the order they'd upheld for billions of years. Throughout the sun, seemingly

344

chaotic atoms were generating light at a highly reliable rate. The sun and moon and Earth were the ceremonial masks of atoms, and the moonlight advancing toward me was their smile. The sound of Lava Falls came from unseen hands pressing piano keys.

I felt it then, the music of the spheres within myself. I was another sphere of harmony upon harmony. I felt my atoms strong and swirling, and on them rested cells of many kinds, full of order yet swirling, and they magnified themselves into organs and bones and heartbeats and muscles and senses and consciousness, orbit upon orbit of harmony, all of it a further refinement of the solar system. Somehow those harmonies had decided that I should be born. Somehow.

The moon remained hidden behind the cliffs but its halo was growing. The moonlight flowed up Lava Falls. I stood waiting to receive it. The moonlight engulfed me.

When I awoke well before dawn, the moon was downcanyon and the entire rapid was lit up, far more luminous than anything else in the canyon. The waves glowed like ghosts.

I wondered how Ancestral Puebloans a thousand years ago perceived Lava Falls. They had indeed been here, for both upstream and downstream of the rapid were a few simple petroglyphs, black lava erupted as faces, faces looking at the rapid, faces now glowing like ghosts.

Before humans emerged from the Sipapu and Ribbon Falls into this world, when they lived in the underworlds without sunlight or moonlight, it was so dim that humans could barely see themselves, and perhaps they didn't even cast shadows. In the emerged, Grand Canyon world, humans suddenly had light by which to recognize themselves. But perhaps it was the moonlight and shadows that offered a truer recognition. With brilliant shadows the cliffs outlined human shapes and motions, emphasized not personal details but simply presence, a strange presence in a stone universe. Was this why shadows were materialized as ravens, allowing the sky to touch itself?

Shadows are part of the magic of Grand Canyon sunsets. At Lava Falls I experienced especially magical sunsets, for the sun was aligned perfectly with the cliffs and sank right into the canyon, an ultimate Stonehenge experience. I watched the canyon revealing a

CANYON AND COSMOS

deeper identity, its cliff faces sharpening, its colors brightening and changing and flowing. Searchlights played through the buttes and cliffs. The canyon glowed with greater intensity. Shadows emerged and grew and grew deeper, gently but relentlessly taking over the canyon. I watched shadows moving upstream, tasting and swallowing Lava Falls, engulfing me, a raven wing embracing me, recognizing my shadow as itself.

Sunsets seem to rebuke humans for letting another day go by without noticing the magic of light and color and the whole world.

My final Lava Falls sunset was surrealistically brilliant with colors. It seemed one of the best sunsets I had ever seen anywhere. I told myself I should remember this sunset for a long time. I told myself that someday, when I was dying, this might be a good thing to remember. I knew this wouldn't really happen, knew that even the strongest experiences fade away.

Even as I was marveling at this sunset, I was feeling discouraged. This was the end of my week at Lava Falls. I had come here hoping that the intensity of Lava Falls would stir a greater intensity in me, as it had when I'd been about to plunge into it. But I hadn't come close to that. I'd had only some flickers and blushes of that. Lava Falls was doing its best to stimulate my senses, but standing safely on shore didn't come close to the full body and full mind engagement of throwing your life into the grip of such intensity. For a week I had seen one thing vividly: the dullness of the human mind. If Lava Falls could not break through and stir my mind, what could? Would I ever experience such intensity again? Was I going to pass through the rest of my life in dullness?

Feeling dim and discouraged, I watched the stars emerge. I watched the stars dimly. Ravens might not have the heat to ignite visions, but what excuses do the stars have? They should set us ablaze. I watched the stars dimly. I felt sorry for myself. I felt sorry for the stars. They deserved better. If the universe depends on Earth eyes and minds to see itself, the universe is being cheated. Is the universe doomed to pass from beginning to end as only a rumor? So much light, so little seeing.

346

THE RAVEN

When I awoke, I lingered in dream twilight zone surrealism, lingered in reality. At this moment, in this strangeness, it seemed plausible that I was only a dream being dreamed by the universe, by empty space seeking to fill itself, confused space seeking coherence, that I was another black monolith surrounded by mysterious black monoliths. I was a shaman raven dreaming she is pursuing herself deeper and unreachably deeper into the sacred erupting heart of mother night.

The river was beginning to drop, and Lava Falls was calming down. As I was filtering my morning water for my hike out, I knew that this was my final Lava Falls water and that soon I'd return to drinking city water. I thought of how I'd been drinking from Lava Falls for a week now. Maybe seven gallons of Lava Falls. Seven gallons that never made it to the bottom of the rapid and continued downstream. By now Lava Falls water was a large portion of my body, maybe fifty pounds of me. This wasn't water steering and drenching a kayaker from the outside. This was Lava Falls become human. I felt it then, Lava Falls flowing through me. I felt its love of falling. I felt its persistence in my heartbeat. I heard its thunder in my breath and even in my silence. I felt its strength in my grip. I felt it becoming waves of consciousness. I felt it trying to erode my human boundaries, trying to offer me its power, trying to stir my senses from within, trying to carry me away, a god taking possession of me.

Perhaps this was why it finally occurred to me, after a week, that some of the water in the river had come from the Sipapu and Ribbon Falls. It had emerged from the Hopi and Zuni portals of emergence and held the powers of creation.

From dark underground, from their mineral volcanoes, the water had bubbled and splashed forth, sparkling with sunlight and reflections of the upper world, flowing down the Little Colorado River or Bright Angel Creek for several miles, joining the Colorado River. It diffused into the river, becoming indistinguishable from other water. It had flowed through the house of creation, reflecting eons of sedimentation and living and dying and evolution and tectonics. It had flowed through the house of the gods, past the shrines through which creation had tried to identify itself, mingling with waters flowing down from the Tower of Ra and Buddha Temple. It had revealed the duality of creation and destruction by peeling grains off rocks and carrying them toward future rocks. It had flowed through dozens of rapids without worry. From the Sipapu it had flowed 120 miles, and from Ribbon Falls it had flowed 95 miles, to arrive at Lava Falls. But just when it was trying to unleash its worst hidden chaos, it was diverted into my water bottles, and into me.

The waters of creation had been flowing into me for a week now, infiltrating me, becoming me, baptizing me from within. Traveling through my veins reminded them of the underground passageways to the Sipapu and Ribbon Falls. My bones were a further calcium dome, my eyes a further portal to the shining world. The waters of creation could now see what they had created, could speak with more than ripples.

I quieted and searched for the waters of creation within me, waiting for them to reveal their presence and power. I thought I heard the bubbling and splashing with which they had emerged. I seemed to feel them. They were feeling right at home. My body had been their body all along. Now they were trying to energize my mind, to turn mere sight into vision.

Some memories lit up. I remembered how at every Lava Falls dawn, birds started singing—the river started singing through them. The bird-river started flying. Out of the retreating night, ravens flew, the night flew with obsidian-black wings. The night flew bravely into the sunlit world. From the underworld, life emerged yet again. The night was too restless to remain hiding in darkness. Out of the night the ravens flew in a ceremonial reenactment of how the universe had indeed congealed into ravens. The blackness had cast a black rainbow. The blindness had sprouted faces and eyes. The vast black wings pushing the galaxies outward had also swirled inward into wings proclaiming the universe's deep genius and relentless questing.

The waters of creation also informed me that I really didn't need a Sipapu or Ribbon Falls to find the waters of creation. If the waters from the Sipapu and Ribbon Falls were indistinguishable from the rest of the Colorado River, perhaps this was because the entire river

THE RAVEN

was the waters of creation, flowing from the Sipapu of the Big Bang. The waters of creation had inhabited me all along, had created me, were energizing my seeing rivers and birds flowing.

Yet this was not the vision I had come here to seek. This was simply a vision of nature, of how everything fits together. I was supposed to be seeing beyond nature, how everything did not fit any logic or necessity. I was supposed to be seeing the ground ungrounded, nature flying inexplicably.

I tried to invoke symbolism that might help me. I thought of the Sipapu as an ingenious symbol of how creation emerges out of strange depths. I thought of how flying ravens seemed to emerge out of nowhere—yet of course this was only the failure of my seeing.

With the waters of creation reclaiming and energizing my mind, I supposed I should finally be able to see primordial creation in everything, with full intensity. I should be able to see a raven as complete mystery. Cosmic mystery should be flowing through me, seizing me, seeing itself clearly, speaking its true name at last.

I looked around for ravens. I stood there and looked back and forth and up and down, at the sky and boulders and cliffs. I turned around and looked around. I waited. I waited, but there were no ravens anywhere. I couldn't wait forever. It was as if the ravens were trying to frustrate me further and finally. It was as if the ravens had all gone into hiding. It was as if this was their final trickster trick in a week of hiding from me.

And with that thought, the waters of creation answered: *Yes, you've finally figured it out. They have been hiding from you, in one way or another, for seven days, the time it took to create the world. These are shaman creator ravens and they've been sending you a message, concealed but perfectly obvious.*

I thought about this awhile. The ravens were sending me a message by hiding? What was the logic of this?

But of course. This wasn't about logic. The universe isn't logical. Creation is a mystery and its origin will remain hidden from us. We can yearn into the universe with the greatest telescopes and theologies and poetry, but its ultimates will remain hidden. Even its mystery is very hard to see.

Of course. The ravens had been hiding from me to teach me that the most important things are hidden from us and will always remain hidden from us.

It was time for me to go.

It was time for me to be here.

Acknowledgments

When Loren Eiseley died in 1977, his longtime personal assistant, Caroline Werkley, returned to her small hometown in Missouri, where she found no one who appreciated Eiseley, how he was the first literary nature writer to make evolutionary time fully real and personal and poetic. I lived near Caroline's hometown, and she was delighted to have someone with whom she could share her history and enthusiasms, and she mentored my writing. She started writing a book about "the hollow," a ravine that as a child she'd found both enchanted and ominous. She was filling the hollow with her favorite characters from mythology and literature and science and history, and together they were having great adventures and profound conversations. Caroline loved trees, so of course the hollow held Yggdrasill, the Viking world tree. To the hollow came scientists like Charles Darwin, writers from Shakespeare to Keats to Ray Bradbury, mythic characters out of Shakespeare and Camelot, and of course Loren Eiseley, who would reveal the hollow's evolutionary secrets.

As I worked on this book, it occurred to me that I was writing a grander version of Caroline's hollow book.

Another free spirit who encouraged me for years was Terra Waters.

I greatly thank the University of Nevada Press and its director JoAnne Banducci and acquisitions editor Curtis Vickers for being bold enough to take on an unusual book. The press's Caddie Dufurrena helped midwife this book into the world. I was fortunate this book was copyedited by John Mulvihill; he raised many good questions and suggested many good answers.

Serena Supplee's painting *Downstream Dream* was perfect for

ACKNOWLEDGMENTS

the cover and themes of this book. Serena is a longtime Colorado River raft guide and perceives and expresses the wonders of rivers and canyons better than anyone.

An earlier version of the "Canyon Sky" section was published in the journal *Pilgrimage*.

Other literary friends who offered encouragement along the way were Richard Berry, editor of *Astronomy* magazine, who published many of my early essays, Pat Garber, Sue Wicklund, Kay McLean, and Rebecca Lawton.

The Grand Canyon takes a lot of time and effort to get to know well, and among those who guided my explorations were Kim Besom, Colleen Hyde, Mike Quinn, Wayne and Helen Ranney, Larry Stevens, Jan Balsom, Tom Martin, Hazel Clark, Tom Bean, Brad Dimock, Richard Quartaroli, Earle Spamer, Alan Petersen, Alfredo Conde, Sheri Curtis, Stew Fritts, Karen Greig, Nancy Brian, Patrick Conley, Susie Bragg, Gale Dom, Stewart Aitchison, Tom Myers, Lynn Hamilton, Mike Anderson, Ron and Pat Brown, Bill Bishop, Dan Cassidy, Michael Collier, Rose Houk, and Mary Williams. My river friends included Linda Jalbert, Lynn Myers, and Patti Lockwood. The Stilley family gave me a great place to write. At Lowell Observatory, Kevin Schindler, Antoinette Beiser, and Michael and Karen Kitt helped keep my eyes aiming upward as well as downward.

About the Author

DON LAGO sought out the Grand Canyon because it seemed the perfect Walden Pond for exploring the place of humans in the grand scheme of things. As happens with many people, the canyon grabbed his imagination and refused to let go. He has spent thirty years exploring the canyon, kayaking it six times, backpacking it some seventy-five times, and doing archaeology and river research for the National Park Service. He has done extensive research into Grand Canyon history, resulting in three books. *The Powell Expedition: New Discoveries* offered twenty years of research and new findings about the 1869 river expedition of John Wesley Powell. *Grand Canyon: A History of a Natural Wonder and National Park* is a cultural human history of the park. *Canyon of Dreams: Stories from Grand Canyon History* relates largely unknown events from national park history. He has also published three books of literary nature and astronomy writing, including *Where the Sky Touched the Earth: The Cosmological Landscapes of the Southwest*. He lives in a cabin in the pine forest outside Flagstaff, Arizona, a town whose culture combines the Grand Canyon, astronomy, and Native American culture.